TOURISM AND RECREATION

access to geography

TOURISM AND RECREATION

Jane Dove

Hodder & Stoughton

A MEMBER OF THE HODDER HEADLINE GROUP

Acknowledgements
The publishers would like to thank the following individuals, institutions and companies for permission to reproduce copyright illustrations in this book:

© National Trust, page 33; © 'Visits to Visitor Attractions', Statutory Tourist Boards of England, Northern Ireland, Scotland and Wales from www.staruk.org.uk, page 9; © World Tourism Organization, page 16.

Every effort has been made to trace and acknowledge ownership of copyright. The publishers will be glad to make suitable arrangements with any copyright holders whom it has not been possible to contact.

Note about the Internet links in the book. The user should be aware that URLs or web addresses change regularly. Every effort has been made to ensure the accuracy of the URLs provided in this book on going to press. It is inevitable, however, that some will change. It is sometimes possible to find a relocated web page, by just typing in the address of the home page for a website in the URL window of your browser.

Orders: please contact Bookpoint Ltd, 130 Milton Park, Abingdon, Oxon OX14 4SB. Telephone: (44) 01235 827720. Fax: (44) 01235 400454. Lines are open from 9.00 to 6.00, Monday to Saturday, with a 24 hour message answering service. You can also order through our website www.hodderheadline.co.uk.

British Library Cataloguing in Publication Data
A catalogue record for this title is available from the British Library

ISBN 0 340 812 48 6

First Published 2004
Impression number   10 9 8 7 6 5 4 3 2 1
Year   2010 2009 2008 2007 2006 2005 2004

Cover photo: *The Broads* by A. Michael © Science & Society Picture Library.
Produced by Gray Publishing, Tunbridge Wells, Kent
Printed in Malta for Hodder & Stoughton Educational, a division of Hodder Headline, 338 Euston Road, London NW1 3BH.

Contents

1 Development of Tourism in the UK

1 Defining Tourism, Recreation and Leisure

Recreation is widely recognised as engagement in activities, pursuits and events such as playing golf, cycling or attending a football match. Where the activities are governed by a set of rules, for example competing in a swimming gala, the term 'sport' is often applied. **Tourism** is linked to the concept of travel to places away from the home environment, for the purpose of leisure, holidaymaking, visiting friends and relations (VFR), pilgrimage or business. The maximum length of the stay, as defined by the World Tourist Organisation, is 1 year but no minimum period is stipulated, which avoids the problem of distinguishing 'tourists' from day-visitors. Other organisations have, however, suggested that tourists are people who stay away from home for a minimum of 24 hours.

Leisure is a broader, overarching term that has traditionally been regarded as non-work time. This notion has, however, been challenged on the grounds that it discriminates against women who work in the home and it fails to take account of flexible working practices, which make the distinction between work and non-working time increasingly blurred. The term 'leisure' also means different things to different people and for this reason it has been suggested that perhaps it should be perceived as a state of mind which diffuses through work and play.

There is a considerable amount of overlap between leisure, recreation and tourism. For example, British citizens who play tennis while holidaying in a Spanish resort would be engaging in leisure, recreation and tourism. Attendance at a local football match would be regarded as both recreation and leisure. Some activities can, however, be more narrowly defined, for example business travel would be classified as tourism but not leisure, assuming this term equates with non-work time.

2 The Development of Tourism in the UK from 1700

a) Beginnings: the inland spa and the rise of the coastal resort to 1918

Tourism in the UK began in the 1700s when inland spas and early sea-side resorts became fashionable with a wealthy, social élite. At locations such as Bath and Tunbridge Wells mineral waters were believed to offer cures for a variety of ailments, and these places began to develop from the early 1700s. Soon the wealthy, with no need of the curative waters, were visiting the spa towns for the season to attend concerts and dances in assembly rooms, to shop and prom-enade. The population of Bath grew from 3500 to 13,000 between 1700 and 1760, and reached 33,000 by 1800. Other spas such as Buxton and Harrogate developed later, becoming popular in the 1850s.

At some seaside locations local people probably engaged in sea bathing as a recreational activity prior to the establishment of tourism. The first coastal **resorts** developed as the result of the belief that sea bathing and drinking sea water were beneficial to health. Early resorts such as Brighton, Weymouth and Margate were located within stagecoach travel of London where the rich lived (see Figure 1). Brighton also benefited from royal patronage, becoming a favourite resort with the Prince Regent who later became George IV. In the north of the country the wealthy visited Scarborough on the Yorkshire coast. This resort initially became popular in the 1730s because spa waters emerged onto the beach and later in the century grew as a centre for sea bathing. Smaller resorts in Devon, such as Sidmouth and Dawlish, developed in the late 1700s in response to demand from the local gentry. South-coast resorts were also stimulated at this time because of the Napoleonic wars that prevented foreign travel. Resorts were modelled on inland spas and had libraries, assembly rooms, ballrooms and esplanades. Promenades became resort status symbols and were places where polite society could be seen and exchange news. Piers, which were initially constructed to allow ships to unload in the 1830s, later became popular places for people to promenade. Numbers visiting resorts, however, remained low because transport by stagecoach was slow, expensive and uncomfortable on the poorly maintained roads.

Moreover, a majority of the population had no paid holiday and were too poor to afford the luxury of an excursion to the coast. From about 1815, seaside resorts began to become popular with a new emerging middle class. Steamship travel developed in the early 1800s stimulating the growth of resorts along the north Kent coast. For example the population of Margate grew from 22,000 to over 88,000 between 1815 and 1840 partly as the result of the development of a fast, safe steamship service from London. Meanwhile on the Clyde, the steamship encouraged travel between Glasgow and resorts such as Greenock. The development of passenger trains from 1830s helped to revive resorts with poor connections and stimulate the growth of others. Most tourists in the 1830–1840s were, however, still drawn from the middle and upper classes.

Resorts in the 1850s were still concentrated along the south coast because of their proximity to London. Many were sheltered from northerly winds, had long hours of sunshine and long, sandy or pebble beaches backed by gently sloping cliffs. Local entrepreneurs, aware of the potential of developing these resorts, invested in hotels and infrastructure, which in turn stimulated more tourist trade (see case study of Bournemouth).

The middle classes of Bristol visited Weston-super-Mare and Clevedon, while in northern Britain industrialisation was creating a prosperous middle class who were travelling to newly developed resorts such as Llandudno in North Wales, Southport in Lancashire and Scarborough in Yorkshire. As the resorts developed, the construction of hotels, boarding houses and places of entertainment further helped to stimulate demand and encourage growth.

By 1870 the concept of a holiday at the seaside was filtering down to the less affluent classes. Cheap rail travel permitted the working classes to travel from the cotton mills of Blackburn and Burnley in Lancashire to the nearby resorts of Blackpool and Morecambe. Regular and reliable employment in the mills, a reduction in the working week and the Bank Holiday Act of 1871 also encouraged the working class to travel to the seaside. Additionally, 'Friendly Societies' encouraged workers to save for a holiday. Workers in the woollen mills of Bradford and Leeds in Yorkshire soon followed suit and travelled to Scarborough for their holidays.

By 1900 Blackpool was growing faster than Southport largely because it developed a range of attractions including theatres, ballrooms and a circus. On the east coast, Skegness and Cleethorpes were created when the railways connected them with urban areas such as Nottingham. As resorts developed they became more specialised in the markets they served, for example Blackpool and Southend focused on the working class, while Bournemouth and Torquay on the south coast, Llandudno in north Wales and Lytham St Anne's in Lancashire were all visited by the middle class.

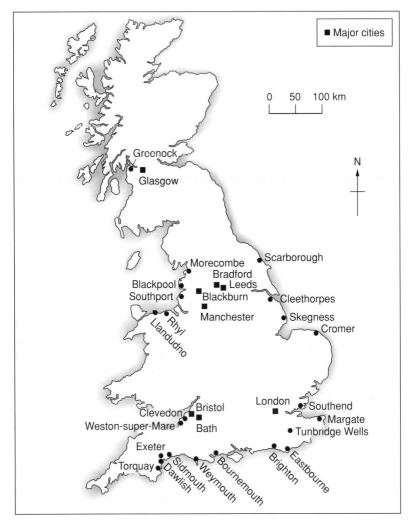

Major cities ■

0 50 100 km

N

Greenock
Glasgow

Scarborough
Morecombe
Bradford
Blackpool Leeds
Southport Blackburn
Cleethorpes
Rhyl
Llandudno Manchester
Skegness
Cromer

London
Clevedon Bristol Southend
Weston-super-Mare Bath Margate
Tunbridge Wells
Exeter
Torquay Sidmouth Weymouth Brighton Eastbourne
Dawlish Bournemouth

Figure 1 Location of inland spas and seaside resorts

b) Rural tourism to 1918

Prior to the 1700s, rural landscapes were largely perceived as wild, dangerous and uncivilised places. Artists and writers were instrumental in encouraging early visitors to travel to rural areas such as the Lake District in the 1740s, although the hills were viewed from ground level rather than climbed. Guidebooks on the Lake District began to appear in the 1780s and tourists started to venture onto the mountains. Interest in the Lake District and the Wye Valley was further stimulated at this time by the travel writings of William Gilpin.

Later on, artists and writers associated with the Romantic Movement such as Wordsworth promoted the Lake District while Jane Austen extolled the virtues of the Peak District in novels such as *Pride and Prejudice*. By the mid-1800s, Thomas Cook was hiring special trains for visitors to travel to the Lake District and North Wales. Meanwhile, the novels of Walter Scott helped to romanticise the Scottish Highlands. Scotland became popular with the upper classes as somewhere to hunt red deer and other game. Thomas Cook began offering highland tours to the middle classes in 1846. Queen Victoria's decision to establish a royal residence at Balmoral further popularised Scotland and railway expansion later encouraged the growth of hotels on the west coast at Oban. Towards the end of nineteenth century the urban middle classes began to explore other rural areas such as the Cotswolds.

c) Tourism in the inter-war years (1918–39)

After 1918 seaside resorts became less exclusive, and the middle and upper classes instead travelled abroad. The number of people taking domestic holidays, however, increased between 1918 and 1939 for a number of reasons. Paid holidays became more common and newer, cheaper forms of transport, notably buses, appeared. Railway companies promoted holiday regions, such as the South West, and holiday camps were developed. The first Butlin's Camp was opened at Skegness in 1935, followed others at Filey and Clacton. These provided holidays for skilled workers and the lower middle class, but were beyond the means of the poor.

For many people the seaside resort remained the most popular destination; Blackpool, for example, received 7 million visitors in 1937. In rural areas, coach trips for the working classes helped to open up the countryside. Cycling, which had become popular with the middle classes in the 1880s, now provided the less wealthy with a means of visiting rural areas. Youth hostelling and camping grew in popularity aided by bus and rail transport. Much of the land in the countryside, however, remained privately owned. The public became increasingly frustrated by lack of access, which led to the Kinderscout March over the Derbyshire peat moors just before World War II. This mass trespass hastened the development of National Parks after the war.

d) Tourism from 1945 to the 1960s

Seaside holidays remained very popular after 1945. The National Parks and Access to the Countryside Act of 1949 paved the way for the opening of the Peak District and the Lake District National Parks in 1951. **Domestic tourism** grew rapidly during the 1950s encouraged by more paid holidays, an increase in wages and advertising by the travel industry. People now had higher expectations than before and an annual holiday was regarded as the norm.

Greater car ownership and the flexibility that this brought opened up previously inaccessible areas such as parts of Wales and Cornwall. Caravan and camping sites grew along the coast providing people with cheap and flexible holidays.

e) Decline of the traditional seaside resort and the Butler model, 1970s to the present day

Holidaymaking at the coast reached its peak in 1970 and from then on many traditional British seaside resorts started to decline. For example, Scarborough experienced a 55% reduction in bed spaces between 1978 and 1994, and during the 1980s British seaside resorts lost 21% of their market. Seaside visitors increasingly tended to be day-trippers rather than long-stay tourists and many were drawn from lower socio-economic groups and the elderly.

A number of external and internal factors contributed to the decline of the traditional seaside resort. Externally, resorts faced competition from the cheap, foreign, package holidays to Spanish resorts such as Benidorm where summer sunshine was guaranteed. Later on, UK resorts also faced competition from resorts developing in Greece and Turkey. Holidays in rural Britain also became more popular as better roads improved accessibility to National Parks and Areas of Outstanding Natural Beauty. The development of holiday-village complexes, such as Center Parcs with its all-weather facilities, and the growth of theme parks, such as Alton Towers, have also contributed to the decline of the traditional resort. The growth of urban tourism and the promotion of short-break holidays to cities such as Bath and Paris also adversely affected seaside resorts.

Internally, at the resorts themselves, accommodation had become outdated and in need of modernisation. Many hotels lacked the en-suite facilities which visitors had come to expect elsewhere, and full board was often viewed as unattractive with families who preferred self-catering arrangements. For their part, hoteliers found it difficult to upgrade their facilities in the light of falling incomes. Parking provision was also often inadequate in resorts, many of which date from the Victorian era when there were no cars.

Some resorts, faced with falling revenues and insufficient visitor numbers to sustain facilities and attractions, converted hotel accommodation into offices, retirement and nursing homes, and hostels for the unemployed. Others sought to invest, diversify or reposition themselves in the market and improve their environments and services. With the help of local authority grants and private investment, relatively large resorts such as Torquay and Blackpool were able to upgrade hotels, convert outdoor pools to indoor leisure centres, construct shopping malls, improve car park provision and develop self-catering accommodation. Restoration of piers, promenades, bandstands and railway stations, refurbishment of visitor centres, and the raising of beach and water quality to meet EU Blue Flag standards have all helped to improve the physical appearance of resorts. At Brighton, for example, the old town has been pedestrianised and the Royal Pavilion upgraded. Plans to restore the West Pier to its '1920s' glory, paid for by property developers and the Heritage Lottery Fund, may go ahead despite a recent fire. Resorts such as Brighton and Bournemouth (see case study) have diversified into business tourism by building conference centres. Others, such as Torquay, have attempted to reposition themselves by moving away from the mass market towards higher spending visitors.

The rise and fall of the of the traditional UK seaside resort can be equated with the **Butler resort area life-cycle model**. In the model, Butler (1980) suggests that resorts pass through six stages based on changes in visitor numbers over time. A resort in Stage 1, or the 'exploration' phase, receives only a small number of visitors and has few tourist facilities. Genuine contacts between the tourists and locals develop and tourism has little environmental, cultural or economic impact. In Stage 2, the 'involvement phase', locals provide a few facilities for tourists but numbers of visitors are still small. A tourist season, and market area from which the visitors are drawn, can be identified. This stage would be typified by Brighton in 1750. In Stage 3, or the 'development' stage, large numbers of visitors arrive and control of facilities now passes to external organisations. Increasing tensions develop between tourists and local residents. In Stage 4, the 'consolidation phase', tourism becomes a major contributor to the resort economy. Visitor numbers reach their maximum and there are signs that some older facilities need upgrading. This stage is typical of British resorts in the 1950s and early 1960s. In stage 5, or the 'stagnation phase', no further increase in visitor numbers occurs, facilities are in need of refurbishment and the resort becomes unfashionable. Five possible scenarios are envisaged at Stage 6, or the post-stagnation phase, namely: immediate decline, decline, stabilization, reduced growth and rejuvenation. Many UK resorts are in the post-stagnation stage today. As previously described, restructuring can take a number of forms. If restructuring occurs it is assumed that existing facilities will need to be upgraded and new ideas introduced if the resort is to remain competitive.

The model can be applied to a variety of seaside resorts, for example those developed in Spain in the 1960s, as well as other types of holiday centre. Critics of the model argue that it fails to acknowledge that different resorts grow at different rates dependent on factors such as accessibility, planning restrictions, market areas and government influence. They suggest that not all resorts necessarily pass through all of the stages identified, and length of stay and spending power is ignored in any calculation of visitor numbers. They argue that tourism in the early stages may lead to improvements to the environment rather than degradation. Furthermore, large companies may be able to protect the resort's market share in the later stages, and the model ignores the impact of surrounding resorts.

f) Changes in the last 30 years

In rural areas tourism has been promoted as farming has become less profitable. Farmers have attempted to diversify by, for example, converting barns into visitor accommodation and offering fishing and horse riding. Other rural attractions, which have developed in rural areas, include wildlife parks, for example Cotswold Wildlife Park in Oxfordshire, and heritage steam railways, such as the Watercress Line in Hampshire. More active types of recreational activity have developed, such as cycling and white-water rafting. Walking in the countryside has become more popular. Membership of organisations such as the National Trust and the Royal Society for the Protection of Birds has grown. Second-home ownership has also increased in rural areas (see Chapter 5 for more detail of rural changes). Holiday villages, such as Center Parcs at Elvedon, and theme parks, such as Alton Towers, have also been built in rural areas (see Chapter 3).

Cities such as York and Glasgow have witnessed a growth in tourism and have become popular short-break destinations (see Chapter 4). Conference tourism has developed in cities such as Glasgow and Birmingham. Some coastal resorts such as Brighton have been upgraded and reinvented. Industrial heritage has become a major tourist attraction at locations such as Ironbridge in Shropshire. Millennium funding has helped to promote new attractions such as the Eden Project in Cornwall (see Chapter 3). Some derelict dockland sites have been redeveloped into marinas such as St Katherine's Dock in London. Popular paid-admission attractions in 2001 included the British Airways London Eye and the Tower of London, while Blackpool Pleasure Beach and the National Gallery were major free attractions (see Figure 2).

In 2001 the UK received 22 million foreign visitors. Although impressive, numbers were 9% lower than 2000, a figure explained by events such as the outbreak of foot-and-mouth disease in March and the September 11 terrorist attack in New York. In an effort to increase number of visitors, a new organisation called VisitBritain was formed in March 2003 to promote the UK abroad and encourage more people in Britain to holiday in this country. The importance of tourism to the UK economy is underlined by the fact that it creates more than £70 billion and supports 128,000 small businesses each year.

One of the aims of VisitBritain is to encourage adults who are not tied to school holidays to take short-break holidays during the spring and the autumn, when attractions and resorts are less crowded. Another aim is to promote sustainable tourism in areas which are currently overcrowded. The task ahead for UK tourism is, however,

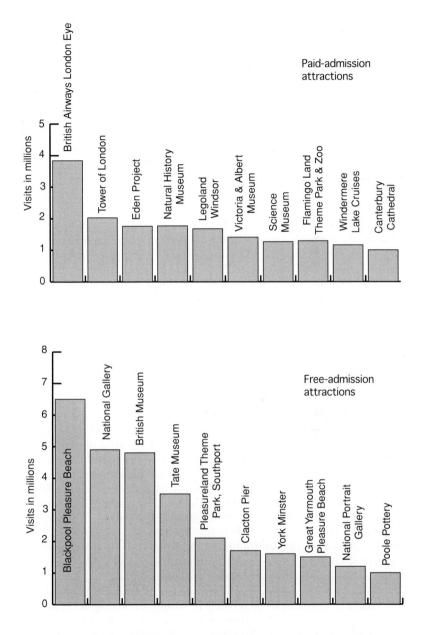

Figure 2 Major paid-admission attractions and major free-admission attractions in 2001 (Source: www.star.org.uk)

difficult because there are already nine English regional tourist boards, nine English regional development agencies, several other public and private groups and tourist boards in Wales, Scotland and Northern Ireland also trying to promote domestic tourism and encourage visitors from abroad.

CASE STUDY: BOURNEMOUTH

In the early 1800s Bournemouth consisted of little more than an inn located on heathland between Christchurch and Poole in Dorset. In 1811, a local squire called Tregonwell built a summer house on the site of the now Royal Exeter Hotel, west of a stream called the Bourne. In 1837, Tapps-Gervis, a local landowner, saw the potential of developing Bournemouth into a fashionable Victorian resort for an emerging wealthy middle class. The site had a mild south-coast climate and sandy beaches backed by gently rising cliffs occasionally broken by deep ravines called chines. Pine trees, planted on the heathland by Tregonwell, also produced a scented air, which was believed to provide a cure for tuberculosis. Tapps-Gervis duly built Westover Villas overlooking the Bourne and the Bath Hotel. Visitors to the large villas that overlooked the Bourne between 1840 and 1870 largely came for the winter season.

The population of Bournemouth grew from 695 to 60,000 between 1851 and 1900. The railway, which arrived in 1870, brought middle-class visitors who came in the summer season. The wealthy winter-season visitors continued to stay in the villas, while houses were converted to boarding establishments for the summer visitors at locations such as Boscombe further east along the coast. Between 1890 and 1914 the number of wealthy winter-season visitors declined while the number of summer visitors grew. During this period Bournemouth also became a popular retirement centre. The Winter Gardens, a concert hall that opened in 1877, offered entertainment for the visitors. A theatre and restaurant complex called the Pavilion was constructed in 1929. More recent developments have included the Bournemouth International Centre, opened in 1983, which has a conference and exhibition hall, fitness centre and leisure pool.

Today Bournemouth has a population of 160,000. It attracts 1.3 million tourists per year, 85% of which are domestic. It also receives a further 4.2 million day-visitors. Tourist, direct and indirect expenditure is estimated at £480 million per year.

The Russell Cotes art gallery was restored in 1997 and there are currently plans to upgrade the Recreational Business District by developing the Grade II-listed Pavilion building into an arts and entertainment centre. There are also proposals to upgrade the landscaped gardens adjacent to the Bourne and to improve car-parking facilities and signposting. The Bournemouth International Centre is also to be refurbished and new attractions developed on the Pier (see Photo 1). Funding for these projects is to be provided by the Regional Development Fund and private investment. Further east at Boscombe, proposals are being discussed to upgrade the pier by creating a piazza and pavilion-style entrance building and new viewing platform. There are also plans to create an artificial surf reef to the east of the pier and a sculpture trail through Boscombe Chine gardens.

The layout of Bournemouth can be compared with a typical resort model as developed by Barrett (1958). The model, developed from observations made at a number of English seaside resorts, is based on the concept that land values decline away from the seafront (see Figure 3). Comparing the model with the layout of Bournemouth, it can be seen that expensive hotels, such as the Royal Bath, are located on the seafront providing ready access to the sea and sea views (see Figure 4). Less expensive hotels and bed and breakfast accommodation occur slightly inland where the land is cheaper. Caravan parks, which require large areas of low-cost land, occur around the edges of the urban area. The railway station would have served the hotels and guesthouses built on the east side of the Bourne. Victorian planners were keen to maximise the beauty of the Bourne stream and its valley, and so converted the area into spacious parks and gardens

Photo 1 Bournemouth Pier

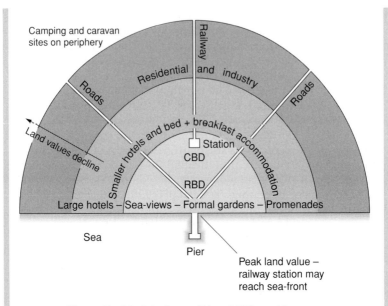

Figure 3   Model of a traditional UK seaside resort

that wind up through the town. The stream and adjacent gardens break the symmetry of the concentric rings in the Barrett model. Pine trees, planted by the Victorians, line many of the residential streets giving the resort a woodland setting. In the model, a recreational business district (RBD) forms in the centre beach-front position and is occupied by food outlets, amusements and theatres. This is well developed at Bournemouth and is focused around the pier and the Bourne outlet. Today, restaurants, a Waterfront Centre and IMAX cinema complex, an Oceanarium, an International Conference Centre and a leisure pool occupy the site. Golf courses, such that in Meyrick Park, and the Bournemouth Tennis Centre are located on the edge of the central area where land values are lower. The university and international airport are located on the periphery of the urban area.

The boundaries of Bournemouth have expanded over time. The settlements of Springbourne and Boscombe to the east of the town were incorporated into Bournemouth in 1876, followed by Winton, Moordown and Southbourne in 1901; Kinson, Ensbury, Wallisdown, Talbot Village and East and West Howe in 1931 and Hengistbury Head in 1932. The effect of this on the model is that satellite leisure centres, for example at Littledown, are embedded in the urban area. Moreover, smaller, secondary recreational business districts have grown up around other piers, such as that at Boscombe.

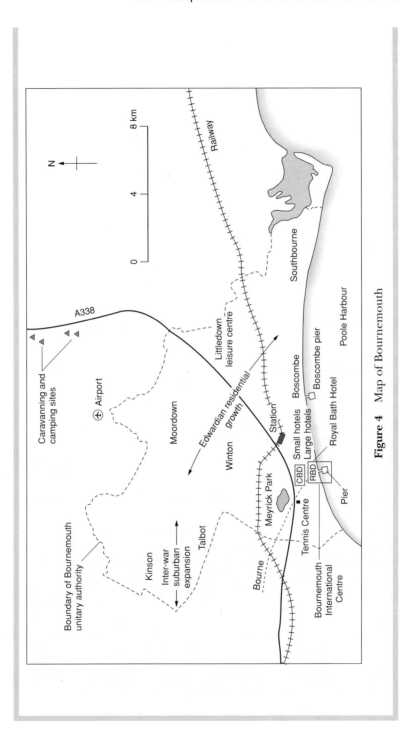

Figure 4 Map of Bournemouth

Summary

1. Tourism in the UK dates from the 1700s when the rich began to visit inland spas and seaside resorts. Visits to rural areas such as the Lake District began in the late 1700s and grew in popularity with the middle and upper class during the 1800s.
2. In the 1800s, steamship travel and later the railways encouraged seaside resorts to grow.
3. The working classes, drawn from industrial towns in the north, began to make day excursions to seaside resorts such as Blackpool from the 1870s.
4. Between 1918 and 1939 improved public transport opened up the countryside, and camping and youth hostelling became popular. Seaside resorts continued to attract many holidaymakers.
5. Greater car ownership in the 1960s opened up previously inaccessible areas to tourism. UK seaside resorts declined after 1970, although some have recently been restructured.
6. Relatively recent developments in domestic tourism include theme parks, farm tourism, heritage tourism and rural holiday-village enclaves.

Questions

1. Why is it difficult to draw distinctions between tourism and recreation?
2. Suggest reasons for the popularity of the major paid-admission attractions and the major free-admission attractions shown in Figure 2.
3. Describe the impact of transport technology on tourism within the UK since 1750.
4. Suggest why traditional seaside resorts in many MEDCs (More Economically Developed Countries) have declined in the last 30 years. Outline attempts to halt this decline.
5. Describe and explain the changes that have occurred in UK tourism in the last 30 years.

2 Growth of International Tourism

KEY WORDS

Tourist-generating country: a country that supplies tourists.
Tourist-destination country: a country that receives visitors.
International arrivals: tourists travelling to a country, which is not normally their place of residence, for more than 24 hours and not longer than a year.
Tourist receipts: money spent by inbound visitors received by the destination country.
Tourism expenditure: total consumption expenditure of outbound visitors for the duration of their stay.
Mass tourism: the concentration of large numbers of tourists in specific locations such seaside and ski resorts often during a clearly defined season. The industry is controlled by large companies who offer all-inclusive packages and is marketed at middle- and lower-income groups. Mass consumption and little differentiation of product are features of mass tourism.

1 Changes in International Tourism Since 1950

International tourists are people who visit a country other than their normal place of residence to go on holiday, to visit friends and relations (VFR), and to conduct business. Changes in tourist flows can be determined from: **international arrivals**, which indicate the popularity of tourist destinations; **tourist receipts**, which reveal the income derived from tourism, and **tourism expenditure**, which provides a guide to high tourist-generating nations (see Figure 5). The figures should, however, be interpreted with care because countries differ in their methods of data collection and some results are based on sample surveys. Tourist arrival data also ignores length of stay, which means that popular short-break destinations are over-represented. Furthermore, residents of small European countries are more likely to cross national borders when they travel than people living in larger places such as the USA, a fact that partly explains the dominance of Europe in international arrivals in the last 50 years.

Before the 1950s, high costs and limited means of transport restricted foreign travel for the majority. Since then the number of international arrivals has risen considerably from 25.0 to 696.7 million between 1950 and 2000. Growth has been continuous and steady over this period, with only minor fluctuations brought about by events such as the oil crisis in the 1970s, the world recession in 1980s and the Gulf War in 1991. In 2001, a general downturn in the world economy and the terrorist attack on the World Trade Center in New York

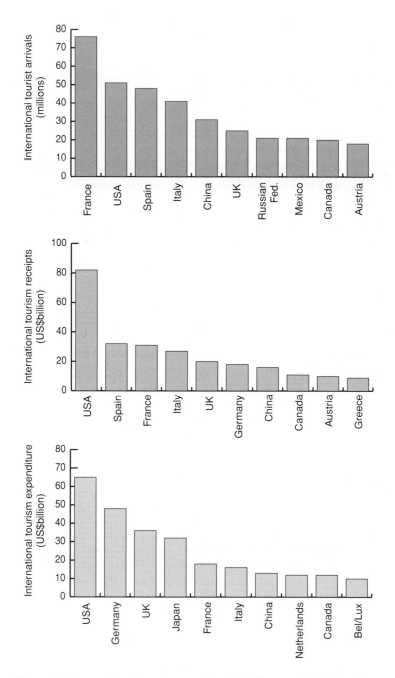

Figure 5 International tourism statistics 2000 (Source: World Tourism Organisation)

caused international arrivals to fall globally (see Section 3 in this chapter), but they rose again to reach nearly 715 million in 2002. Tourism is a major contributor to the economies of many countries worldwide and generated US$474 billion in tourist receipts in 2000. More Economically Developed Countries (MEDCs) have been the main **tourist-generating countries** throughout the last 50 years. High tourist spenders in 2000 included the USA, Germany, UK and Japan (see Figure 5). High tourist flows occur between MEDCs, such as between Canada and the USA, the USA and Europe, and between Japan and the USA/Europe. Large tourist flows also occur between MEDCs and accessible destinations within Less Economically Developed Countries (LEDCs).

Europe has remained the largest tourist-receiving region in the last 50 years, although its market share has fallen. In 2000 it received 57.8% of the global total of international arrivals. Some European countries are more attractive to tourists than others, for example Spain is a high receiver (see Figure 5) but a low generator of tourists, whereas the Netherlands is a high generator but a lower receiver. The main tourist movements are from cooler, north European countries such as Germany and the UK to warmer destinations in Spain and Italy. French and Spanish Mediterranean resorts were very popular in the 1960s, while the Greek Islands and Yugoslavia developed as tourist designations in the 1970s, followed by Turkey in the 1980s. Tourism in eastern European countries such as Poland and Estonia has grown since the Eastern Bloc collapsed in 1989–91.

A resort that typically reflects the growth of Spanish tourism is Torremolinos, which is located 14 km from Malaga. Before 1960, Torremolinos was a fishing village, visited by a few wealthy Spaniards, writers and artists. Rapid **mass tourism** resulted in poorly planned, high-rise hotel developments, polluted beaches, little public space, few parking facilities and an outdated sewage system that could not keep pace with growth. These problems together with friction between northern Europeans and locals, crowded bars and streets, drunkenness and petty crime encouraged the rich and the artistic types to leave. Torremolinos can now be said to be in the post-stagnation phase of the Butler model. Local authorities are attempting to upgrade the resort with developments that include the building of a congress and conference centre.

International arrivals to East and Southern Asia and the SW Pacific rose from 1.0 to 16.5% between 1950 and 2000. Thailand, Singapore, Hong Kong, Japan and Australia have become popular, although in many cases growth has been focused at specific resorts within these countries such as Pattaya (see case study) and Phuket in Thailand. Since China adopted a more 'open door policy' in 1978, international arrivals have increased dramatically reaching 31.2 million by 2000. Some 131 million international arrivals, or 18% of the global market, visited South and East Asia and the Pacific in 2002, although the

Photo 2   Tourists in the Forbidden City, Beijing

terrorist attack on Bali in 2002 adversely affected tourism to that country (see Section 3).

Between 1950 and 2000, the percentage of international arrivals visiting the Americas declined from 29.0 to 18.4%. These figures, however, obscure the fact that the USA was the second most popular **tourist destination** in 2000, receiving 50.9 million international arrivals, and was also the country with the highest tourist receipts of US$82 billion (see Figure 5). The US tourist industry was one of the worst affected by the September 11th terrorist attack in 2001, an event that partly contributed to a 6% fall in international arrivals to the USA over 2002.

Tourism to Africa and the Middle East has remained low. International arrivals to Africa increased from 1.5 to 3.9% between 1970 and 2000, while visitors to the Middle East have risen from 1.4 to 3.3% over the same period.

Business tourism, particularly in the last 25 years, has become an increasing part of the international tourism market in countries such as Singapore. Some trips are undertaken to buy and sell goods, others are offered by companies as perks or rewards to their employees. Conference tourism is a lucrative market and there is often intense rivalry between cities to secure this type of business. Capital cities such as Paris, Madrid and Washington are popular business venues.

CASE STUDY: PATTAYA

In the 1940s, Pattaya consisted of a number of fishing communities on the coast of Thailand, 147 km south-east of Bangkok. In the late 1940s second homes began to appear on the seafront owned by wealthy Thais who normally resided in the capital. Later on more bungalows were built and occupied by expatriates who also lived in Bangkok. Road access to the capital improved in the 1960s and the first hotel opened 1964. In the late 1960s the resort became popular with US servicemen who were on leave from the Vietnam War. The resort was marketed internationally in the 1970s and many more hotels were built. Tourism grew rapidly during the 1980s and by 1988 the resort was attracting 2.9 million international visitors. Domestic tourist numbers also increased as more Thais living in Bangkok travelled to Pattaya for a holiday.

Rapid tourist development has resulted in a number of negative environmental impacts. By the mid-1990s, beaches were disfigured by uncontrolled developments, and rubbish and poor wastewater management had resulted in marine pollution. The resort had also become associated with sex tourism, and drugs and crime were also problems. The Thai Government, mindful of the need to protect and nurture the tourist industry, has tried to improve the image of the resort. A new water-treatment plant was installed, beaches were cleaned up and an Exhibition and Convention Centre designed to attract business tourism.

The growth of Pattaya is illustrated by a beach-resort model produced by Smith (1991). In the model, a pre-tourist stage is followed by a period when second homes grow up along the seafront owned by a wealthy local élite. A few 'explorer' type tourists also visit the area. Improved accessibility encourages a hotel to be built, the success of which leads to others being constructed, which displace residential housing from the sea-front. Local businesses develop to serve the needs of the expanding tourist industry. A business core develops and eventually a central business district (CBD) becomes differentiated from the RBD. Rising visitor numbers lead to problems such as overcrowded beaches, water pollution and traffic congestion.

2 Factors Explaining the Growth of International Tourism

A number of factors explain the growth in international travel including changes in transport, wealth, working conditions, education, the travel industry, technology, fashion and political stability.

Aircraft have become faster and can now cover greater distances before refuelling, which has compressed space and time. This initially helped European tourism to develop in the 1960s and later contributed to the growth of holiday destinations further afield in Asia and the Pacific. The introduction of the wide-bodied jet, which can carry large numbers of passengers, has reduced the cost of long-haul travel. Reduced cost has also allowed relatively new sun–sea holiday destinations in Turkey, for example Bodrum, to compete successfully with existing resorts in Spain. Improvements to rail networks, better tunnels through Alpine passes, the development of high-speed trains, notably Eurostar, motorway improvements, autobahns and continental toll roads have also encouraged European tourism.

Industrialisation, initially in Europe, the USA and Japan, and later in Newly Industrialised Countries (NICs) in Asia and the Pacific has created wealth and more disposable incomes for travel. Greater economic prosperity also explains why many people from the former Eastern Bloc countries are beginning to travel abroad, particularly to the Mediterranean. The growth of business tourism reflects the globalisation of the world economy. Older people, many with disposal incomes, who are fit and keen to engage in foreign travel, now form an increasing percentage of the population of MEDCs and NICs. Longer and more frequent paid holidays in Europe and NICs have also contributed to increased foreign travel.

Improvements in education have increased public curiosity to see more distant, exotic locations, such as the Brazilian rain forests and Antarctica. A desire to see and learn about different cultures, rather than spending 2 weeks on a beach, has encouraged tourism to countries such as China and Egypt. Exotic localities such as the Galapagos Islands have been promoted by ecotourism. Foreign-language teaching in schools has given the public a greater confidence to travel abroad.

The travel industry has become more active in promoting and packaging holiday destinations. Spanish resorts partly developed in the 1960s because travel companies developed all-inclusive, low-cost, easy-to-book, package holidays marketed at middle- and lower-income groups. This formula has since been successfully applied to other European, as well as long-haul, destinations. Package tours often include guides, which reduces any anxieties tourists may have over understanding language and customs. Greater availability of hotel accommodation that conforms to certain standards has made the public more willing to travel abroad. The guarantee of good-quality hotels partly explains the increase in tourism to destinations in NICs, such as Phuket in Thailand, while lack of provision has contributed to low visitor numbers to African destinations. The types of accommodation available have also been widened now to include self-catering, apartments and villas. In some cases the travel industry has been at the forefront of developing new attractions, for example Disneyland Paris.

Advances in technology, such as booking through the Internet, have made travel procedures less complex. Simpler booking procedures have contributed to the success of low-cost airlines such as Ryanair. Medical insurance provision, improved communication links with home and travellers cheques have all helped to make the traveller feel more secure.

Fashion has also influenced the growth and decline of tourist locations. For example, mainland Spain and the Balearic Islands were popular in the 1960s. However, as visitor numbers grew, concerns about over-crowding encouraged some tourists to take their holidays in lesser-known localities in the eastern Mediterranean. The fall in the popularity of Spain partly explains the decline in European tourism arrivals over the last 50 years. The travel industry, sensitive to the concept that foreign holidays are perceived by some as status symbols, has been at the forefront of promoting, and in some cases determining, fashionable holiday destinations.

Post-war political stability helped European tourism to grow in the 1960s. More recently the relaxation of border controls has made travel within the EU easier. The fall of communism has led to increased tourism to and from former Soviet Bloc countries. In contrast, travel companies have been reluctant to invest in politically unstable areas, which partly explains why tourism in much of Africa and the Middle East has remained low. There is further discussion of the impacts of political stability on tourism in the next section.

Future changes in the patterns of international tourism are likely to be influenced by factors such as new forms of transport and the development of LEDCs. Concern that video-conferencing might undermine the business sector appears to be unfounded because 'face-to-face' contacts are still preferred by managers of transnational companies. World events and political changes will continue to cause minor fluctuations in growth.

3 Unforeseen Events and Political Instability and their Impacts on Tourism

Freak weather, political instability and changes in the economy can all discourage foreign travel. Tourists are uncommitted until they have fully paid for package holidays, and when unforeseen events arise they often cancel their plans, or switch to alternative destinations offering similar facilities. In countries where holidays are cancelled some local people can find themselves in debt because they have invested in the provision of services and facilities before the new season arrives. They also lose the opportunity to make a profit that would carry them through until the next season begins. Governments lose foreign earnings and may have to divert funds away from schools and hospitals in order to support a recovery in the tourist economy. In contrast, countries benefiting from cancellations receive a boost to foreign earnings. The triggers responsible for the volatility in the tourist industry are explored in more detail below.

a) Natural hazards

Tropical cyclones, earthquakes and volcanic eruptions can all adversely affect the tourist industry. Visitors fear to travel, hotels, roads and railways, and electricity and water supplies are often destroyed or damaged, and crops lost. For example, in October 1988, Hurricane Gilbert swept across the whole length of the island of Jamaica devastating the tourist industry. More recently, river flooding in Eastern Europe in the summer of 2002 discouraged visitors from travelling to Prague. The 2001, UK foot-and-mouth outbreak badly affected the rural economies of north and west Britain (see case study). The severe acute respiratory syndrome (SARS) epidemic in March 2003 had a detrimental impact on tourism to China and other parts of South-east Asia, although tourism to these countries soon recovered.

CASE STUDY: UK FOOT-AND-MOUTH DISEASE AND TOURISM 2001

Foot-and-mouth disease became a serious outbreak in March 2001 just as rural areas were preparing to receive Easter visitors. The Lake District and the south-west of England were the most, and the south-east the least, affected areas. Footpaths were closed and sporting fixtures such as horse racing were cancelled. Visits to farm attractions fell by 25%, while trips to country parks declined by 6%. Numbers visiting historic properties fell by 7% and excursions to wildlife attractions declined by 4%. The Government was criticised initially for falsely conveying the impression that all of the countryside was out-of-bounds, which further exacerbated the loss in trade for hoteliers and innkeepers. In contrast, some attractions became more popular, for example visits to Legoland at Windsor rose by 9.5%. Visits to Kew Gardens in south-west London also increased by 15%, partly because nearby Richmond Park, with its large deer herd, was closed. Visits to theme parks increased by 4% and trips to gardens by 3%. The retail sector also reported an increase in trade as more people switched to shopping as a leisure activity.

b) Political instability and terrorism

A civil war or a military coup can lead to volatility in the tourist industry. For example, in the late summer of 1994 a bloodless military coup in The Gambia removed the president from office. In November of that year, at the start of the tourist season, the UK Foreign and Commonwealth Office (FCO) advised visitors against travelling to The Gambia, a popular holiday location that in the previous year had received 90,000 arrivals. As a result of this action, 1000 jobs were immediately lost; eight out of the 17 hotels closed and those remaining open operated at less than 50% of their capacity. Craftsmen, who had made souvenirs ready for the tourist season, and taxi drivers, who had ensured their vehicles were in a good state of repair, found themselves in debt. Travellers who ignored FCO advice and took holidays in The Gambia that winter reported no trouble. The following March, just as the tourist season ended, the FCO announced that it was safe for visitors to return to the country. Critics have suggested that the FCO should not have issued a travel embargo to persuade the military government to return to democratic rule. By 1996, political stability was restored and tourist numbers began to rise again.

More recent examples of the effects of political action on tourism can be found in Myanmar and Nepal in South-east Asia. In Myanmar the military government recently relaxed visa regulations and improved infrastructure in an attempt to boost tourist foreign earnings and to win international approval for its regime. Despite these initiatives few visitors are still willing to travel to a country where most of the population receive little benefit from tourism and where profits are taken by a ruling military élite or foreign workers. In Nepal the massacre of the Nepalese royal family in June 2001, and an escalation of the civil war with Maoist revolutionaries in 2002, has resulted in a 70% fall in the numbers of mountaineers, trekkers and holidaymakers visiting Nepal. Nepal usually receives about half a million visitors per year who contribute about US$160 million in foreign earnings. The industry employs 200,000 Nepalese in a country where 40% of the 23 million are living in poverty. In an effort to attract tourists back, the government has opened many new mountain peaks of different heights to climbers, although Everest remains the goal for many. The 2003 celebrations to mark the first ascent of Everest encouraged many climbers to visit the area and there were reports of overcrowding at base camp.

One-off events can trigger instability in the tourist market, but recovery is often swift. In 1978, the Chinese government announced a more 'open door policy' and travel restrictions to China were lifted. As a consequence of this visitor numbers quickly rose to reach 1.7 million by 1987. The brutal suppression of student demonstrations in Tian'anmen Square in June 1989, however, resulted in 23% fall in international arrivals in that year. The impact was, however, short-lived and visitor numbers soon recovered. By 2001, international arrivals reached 33.2 million, the fifth largest in the world and tourist receipts totalled US$17.8 billion.

Acts of terrorism have recently highlighted the volatility of the tourist industry. The attack on the World Trade Center in New York in September 2001, together with a slow down in the global economy, resulted in 4 million fewer, or a 0.6% fall in, international arrivals worldwide in 2001 compared with 2000. Foreign travel to the USA was particularly affected and visitor numbers fell by 20%. Fewer international tourists visited the UK, and visitor numbers were down by 12% at Westminster Abbey, 20% at the Tower of London and 16% at Tate Britain.

In 2002, an attack on a synagogue in Djerba adversely affected tourism to Tunisia, while bombs exploded in streets crowded with tourists in a nightclub area of Kuta, a popular resort on the south-west coast of Bali in Indonesia. Many of the 187 who died were young, backpacking, nightclubbing Australians for whom Bali was their 'Costa del Sol'. The resort was also popular with British holidaymakers who travelled from Singapore and also visited the nearby island of Lombok. The attack has raised fears that other holiday destinations that are popular with Westerners, such as Phuket in Thailand and Langkawi in Malaysia, may also become terrorist targets.

Tourists are also often unwilling to travel to countries with undesirable political regimes, such as Zimbabwe. In contrast, countries recovering from a period of political instability can appeal to travellers because they offer opportunities to explore landscapes, which have previously been out of bounds. Governments can actively help to tourism to recover. For example, after 30 years of troubles in Northern Ireland, a National Resource Rural Initiative has been established to promote rural tourism. It has a funding of £15 million and is managed by the Northern Ireland Tourist Board, the Department for Agriculture and Rural Development and the Environment and Heritage Service. Resources are to be focused on areas designated as Areas of Outstanding Natural Beauty and/or Environmentally Sensitive Areas including The Antrim Coast and Glens, The Mournes, The Sperrins, Fermanagh and South Armagh. These areas will receive funding to improve infrastructure, skills and facilities to develop sustainable tourism.

Summary

1. International tourism has increased significantly since 1950. Europe has remained the largest tourist-generating area in the last 50 years, but its market share has fallen.
2. Visits to South-east Asia and the Pacific have increased significantly since 1970, but those to Africa and the Middle East have remained low.
3. Factors accounting for the increase in international tourism include advances in transport and technology, greater affluence, political stability, fashion, media and travel company promotion, and improvements in education.
4. Tourism is a volatile industry that quickly responds to changes in demand.

Questions

Short answer

1. Distinguish between:
   (a) international and domestic tourism,
   (b) tourism expenditure and tourism receipts.
2. Study Figure 5, which shows international tourism statistics for 2000.
   (a) Suggest why France, Spain and Italy are very popular tourist destinations.
   (b) Identify countries that are: (i) net earners; and (ii) net spenders.
   (c) Suggest why Greece appears as a top tourism earner but does not feature as a top tourism spender.
3. Describe and explain the changes in international tourism in the last 50 years.
4. Suggest why there are so few LEDCs in the list of top tourist destinations and what they can do to increase visitor numbers.
5. Using examples, describe and explain why tourism is perceived as a volatile industry.

3 Redirecting the Tourist Gaze

KEY WORDS

Heritage tourism: a type of tourism where the core product offered, or the main motivating factor for the visitor, is heritage. It includes historical buildings and events, indigenous cultures, traditional landscapes and wildlife, art, music, food and drink.

Theme park: a leisure complex occupying a large area built around a focused theme offering a range of activities including rides, performances and exhibitions especially for families for an all-inclusive price. Experiences are often a mixture of real and artificial: invented places, magical kingdoms and fantasy lands.

Holiday-village enclave: a type of tourist development, which is physically, economically and socially concentrated and self-contained. It provides a range of indoor and outdoor recreational experiences for families who stay in purpose-built accommodation.

National Trust: an organisation founded in 1895 to preserve and provide public access to buildings of historic or architectural interest and landscapes of natural beauty.

1 New Types of Tourism

A number of new forms of tourism and recreation have emerged in the last 30 years. These include heritage tourism, theme parks and rural holiday villages discussed in this chapter, urban tourism covered in Chapter 4 and ecotourism examined in Chapter 7. The Eden Project in Cornwall (see case study) illustrates some of the factors responsible for the growth of new types of tourist experience. Television documentaries such as *Life on Earth* and teaching about rainforests in schools have encouraged awareness and curiosity about tropical environments. Road improvements have made Cornwall more accessible, and greater affluence and more paid holidays have given the public the time and the means to visit the Project. Technological advances have enabled huge greenhouses to be constructed within a quarry and improvements in communications, notably the Internet, have helped to promote the attraction to a wide audience.

Films and the television adaptations have also helped to redirect the 'tourist gaze' towards new localities. For example, Stamford in Lincolnshire witnessed a 27% increase in visitor numbers in the year after the release of George Eliot's *Middlemarch* which was filmed in the town. Similarly, the small village of Goathland, in the North York Moors, became a popular tourist attraction after it was selected as the setting for the television series *Heartbeat*. The recently released Harry Potter films have also helped to promote tourism in the UK (see case study of Alnwick Castle).

In the USA, visits to a dramatic volcanic plug known as The Devil's Tower increased by 74% when it featured in the film *Close Encounters of the Third Kind*. Visitor numbers increased again 12 years later when the film was released as a video. New Zealand's tourist industry has benefited from the release of *The Fellowship of the Ring*, *The Two Towers* and *The Return of the King*, all part of JRR Tolkien's *The Lord of the Rings* trilogy and filmed on location there. Special 'Hobbit' tours take visitors to locations such as Tongariro National Park, the setting for Mordor.

CASE STUDY: THE EDEN PROJECT

Large greenhouses containing a huge variety of tropical plants, located within a disused china clay quarry near St Austell, have recently become a highly successful Cornish tourist attraction. The 'humid tropics biome', which is over 200 m in length and 50 m high and therefore one of the largest greenhouses in the world, contains plants from the rainforests of South America, West Africa and Indonesia. A second greenhouse, called the 'warm temperate biome' contains plants from the Mediterranean and California. A third 'out-of-doors' biome is made-up of plants associated with the British climate.

The Eden Project, which has received funding from the Millennium Commission, the South West Regional Development Agency and the European Objective One Fund, has cost £88 million to date. Opened in March 2001, the Eden Project attracted 2 million visitors in that year, rather than the 750,000 expected. About 650 people are employed at the site and many more jobs have been created in local hotels and restaurants. The estimated benefit to the Cornish economy so far has been £155 million, a welcome windfall for a county which by national standards is poor.

The success of the Eden Project has prompted plans to build a 'dry zone biome', recreating conditions similar those in the Kalahari Desert, at an estimated cost of £75 million. Despite the benefits to the Cornish economy the Project has not been without its critics, who argue that roads have become congested. Managers of the Eden Project are currently in consultation with coach and rail operators to develop more sustainable transport systems. Currently, a shuttle bus links the Eden Project with St Austell railway station.

CASE STUDY: ALNWICK CASTLE AND HARRY POTTER

Alnwick Castle in Northumberland, home to the Percy family, was built in the eleventh century. Restored in the eighteenth century, it contains paintings by Titian and Canaletto and has been open to the public for many years. The castle, together with adjoining Hulne Park and Abbey, have formed the backdrop to several television and film productions including *Robin Hood Prince of Thieves* (1991) and *Mary Queen of Scots* (1971). Filming has helped to broaden the appeal of the castle from an adult historic attraction to a media entertainment for children and their parents. Following the release in November 2001 of *Harry Potter and the Philosopher's Stone*, which was partly filmed at the castle, visitor numbers that had previously averaged around 52,000 per year, rose to 139,461 in 2002. The rise was, however, also partly caused by the opening of Alnwick Garden adjacent to the castle and money received to promote the attraction given by Northumberland Council to offset the impact of the 2001 foot-and-mouth outbreak.

Many of the visitors to the castle come from overseas, notably from the Netherlands, Germany and the USA. The possibility that filming might be occurring during a visit adds to the attraction's appeal. Business tourism has also increased because companies have been keen to book the castle for corporate events. Hoteliers in Alnwick have also benefited from increased visits to the castle.

2 Heritage Tourism

a) Characteristics of heritage tourism

History and culture can be presented in a variety of ways and attractions can vary from the authentic to the totally non-authentic. Figure 6 illustrates five approaches to the presentation of Viking heritage, which range from the conventional museum to a theme park.

Heritage tourism is a broad term embracing visits made to stately homes, restored mills, docks, canals and railways, cathedrals, churches, art galleries, museums and battlefields, and the homes of writers, poets, composers, artists and politicians. It also includes the re-enactment of historical events, traditional landscapes, wildlife reserves, heritage coastlines, national parks and geological sites.

Examples of recently opened heritage attractions in the UK include the National Maritime Museum at Falmouth in Cornwall and

Viking Theme Park – e.g. 'Viking Land' south-east of Oslo. Outdoor reconstructed quay, ship, market place. Indoor multimedia auditorium. Ride attraction takes visitor on a Viking longship voyage to Vinland. Popular with schools.

Viking Village Reconstruction – e.g. Foteviken, near Malmo, Sweden. Houses, workshops, ship reconstruction based on the Viking settlement at Foteviken Bay. Craft activities – dyed wool, books, cards – income from which supports archaeological work.

The Viking Ships Museum, Oslo 'Conventional, authentic museum.' Three Viking ships c. 834 excavated from Viking graves in the 1800s near Oslofjord. Now preserved and reconstructed in a purpose-built museum. Scholarly publications on sale in bookshop.

Jorvik – Discovery of the remains of Viking York during construction of a shopping centre in 1982. Site of excavation now transformed to exhibition of objects in situ. Together with 'time car ride' which takes visitors back through reconstructed alleys and houses of Viking York together with smells and sounds.

Viking Fairs and Markets – Craft fairs often in rural settings where costumed traders sell Viking replica goods. Opportunities to learn how to build Viking houses and boats. Mock-fights.

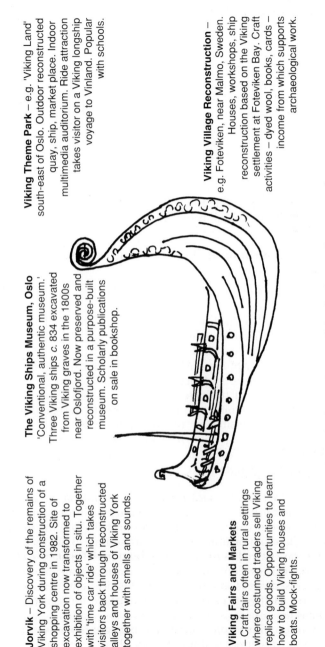

Figure 6   Representing Viking heritage – five approaches

'The Deep', a natural history museum of the sea, located on the banks of the Humber near Hull. This attraction opened in 2002 and received 800,000 visitors in its first year. It cost £45.5 million to construct and was funded by the Millennium Commission, the European Development Fund and Kingston-upon-Hull City Council.

Newly opened heritage attractions on the European mainland include a museum dedicated to the history of the Polish Jews in Warsaw, and a European Park of Volcanism located near Clermont-Ferrand, in France. The aims of 'Vulcania' are to promote a better public knowledge and understanding of volcanism, to boast the economy of the Auvergne and to protect the natural volcanic sites in the Massif Central. Sculptured out of the base of lava flows, much of the attraction is located underground and includes displays of volcanic activity, interactive models and exhibits, shops and restaurants. The attraction opened in 2002 and by September of that year it had received 500,000 visitors. The EU, the French state and the Auvergne Regional Council funded the project.

b) Reasons for growth of heritage tourism, benefits and criticisms

Heritage tourism is not a recent idea, for example in the eighteenth century wealthy young men visited cultural European centres such as Florence and Rome to finish their education as part of 'The Grand Tour'. In the late nineteenth century Thomas Cook organised excursions to historic monuments in Egypt.

There are a number of reasons why heritage tourism has grown in the last 30 years. Television adaptations of novels and documentaries have raised awareness of heritage and make it accessible to a wider audience. Greater car ownership and improvements in road transport have more places accessible in the domestic market. Jet aircraft, cheaper flights and package tours have increased accessibility to international sites. More disposable income and a greater number of paid holidays have given people more opportunity to visit heritage sites. Improvements in education have created a curiosity to learn about history, culture, art and natural landscapes. Deindustrialisation has left a legacy of nineteenth-century warehouses, disused mills, factories, canals, docks and coal mines that provide insights into the industrial past and opportunities to learn something about the lives of the working class as well as the rich (see **National Trust** case study). National and local governments perceive tourism as a means of regenerating depressed areas and conserving historic city centres. The travel industry itself has recognised that heritage tourism can be a lucrative market.

The expansion of heritage tourism has brought benefits including employment in rural areas and inner cities where there are often few other work opportunities (see Chapter 4). Revenues accrued from tourism have been used to restore and maintain buildings often relieving the government of this responsibility. Reconstructions, such as Lascaux II (caves containing prehistoric paintings in France), prevent the originals becoming damaged. Restoration, for example at the Ironbridge Heritage Centre in Shropshire, and reconstruction, for example Beamish in the north-east (see case study), can help to bring history alive, particularly for children.

Critics of heritage tourism argue that some developments cannot be sustainably managed and in time will become overcrowded and deteriorate. Others suggest that heritage tourism leads to artificial branding and creation of stereotypes. For example, the novels of Catherine Cookson and their television dramatisations have led parts of Tyneside in the north-east be repackaged as 'Catherine Cookson Country'. As a consequence of this, visitors perceive Tyneside through the eyes of the author or film director, not as it really is. The mass marketing of heritage produces unauthentic souvenirs and events, which are staged for entertainment and lose their meaning (see Chapter 6). Places may lose their purpose, for example Notre Dame in Paris has become a tourist attraction rather than a place of worship. This is because people come to admire the architectural splendour rather than to pray. Reconstructed heritage can present a cosy, selective and nostalgic view of history, which encourages people to look backwards. Furthermore, the distinction between reconstructed non-authentic attractions and theme parks becomes blurred. Non-authentic attractions may also take custom away from real places. For example, a recent proposal to build a Shakespearian theme park, near Stratford-upon-Avon covering 490 hectares, complete with wattle and daub hovels, a tournament field, gallows, stocks, costumed inhabitants selling their wares and peasants toiling in nearby strip-fields has met with opposition from a Royal Shakespeare Company project that aims to attract more young people to Stratford to see Shakespeare's plays.

CASE STUDY: NATIONAL TRUST

Octavia Hill, Sir Robert Hunter and Harwick Rawnsley established the National Trust in 1895 with the aim of preserving and providing public access to buildings of historic or architectural interest and landscapes of natural beauty. From an initial membership of 100 in 1895, numbers have now increased to over 3 million (see Figure 7). A number of factors have contributed to the particularly steep rise in membership since the 1960s. Greater car ownership has improved accessibility to stately homes in rural locations, as well as to large stretches of the coast and countryside managed by The Trust (Figure 8). Increased

Photo 3 Entrance to Sutton Hoo National Trust exhibition hall

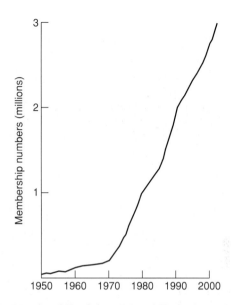

Figure 7   Membership of the National Trust since 1950 (Source: based on figures from National Trust)

urbanisation has encouraged people to visit the countryside during their leisure time as a means of escape. The public have become more sensitive to the need to preserve heritage, and more disposable income has given them the means to patronise The Trust. A larger membership has created greater revenue enabling more buildings and landscapes to be saved for the nation. For example, public donations, together with a £17 million National Heritage Memorial Fund grant, recently helped to secure the Victorian gothic mansion of Tyntesfield near Bristol. Demographic change has meant that there are now older, but still active, people in the population who have the time and ability to visit National Trust sites. Some properties have been used in television dramas, which has made them more popular, for example Lyme Park in Lancashire was the setting for Pemberley in the 1995 filming of *Pride and Prejudice*. The Trust has made a greater range of acquisitions which has broadened its appeal. Recent additions include an Anglo-Saxon burial site at Sutton Hoo in Suffolk (see Photo 3), a nineteenth-century workhouse at Southwell near Nottingham and the home of ex-Beatle Paul McCartney in Liverpool. Servants' quarters in stately homes, such as Petworth in Sussex, have also been renovated to provide glimpses of the lives of the poor as well as the rich. The Trust has encouraged school visits and has developed interactive displays that appeal to the young.

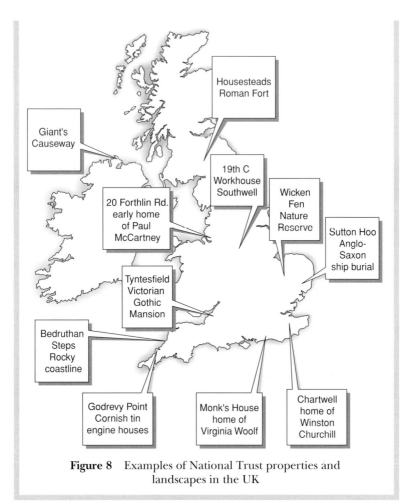

Figure 8   Examples of National Trust properties and landscapes in the UK

CASE STUDY: BEAMISH OPEN AIR MUSEUM

Beamish Open Air Museum is located 14 km south-west of Newcastle-upon-Tyne and covers an area of 740 hectares. Its reconstructed buildings portray the social history of the north-east region in 1825 when the area was largely rural, and again in 1913 by which time heavy industry had developed. The attractions include a colliery village, a mining town, a railway station, a drift mine, a farm and a manor. The museum was opened to the public in 1972 and attracts about 350,000 visitors per year, many of whom come from outside the region. A further 50,000 school children visit the museum annually. The museum employs about 50 people full time and up to 150 more during the peak season. Some 95% of the running costs are met from administration charges, catering, retailing and corporate hires, the remainder comes from a local authority grant. Money for capital projects such as new buildings is funded from public donations and grants from European Regional Development Fund, the English Tourist Board, the Heritage Lottery Fund and the National Heritage Memorial Fund. A committee representing the city, district and county councils administers the museum. Beamish was named the Best UK Attraction for group visits in 2002.

3 Theme Parks and Holiday-village Enclaves

Amusement parks that grew in popularity during the late 1800s were the basis for the development of the **theme park**. Early examples of theme parks, including Disneyland, at Anaheim, Los Angeles, in California which opened in 1955, and Disneyworld in Florida, were located near large urban markets in the warmer parts of the USA. They were an immediate success because they provided safe, controlled, clean environments for leisure, and the public could easily comprehend, label and consume the different themes on offer.

From the USA the theme park concept spread to Europe and later to East Asia. Thorpe Park was the first theme park to open in the UK, developed around disused gravel pits 30 km west of London. Since then many others have been built including Legoland near Windsor, Flamingo Land at Kirby Misperton in Yorkshire and Flambard's Village in Cornwall. The largest theme park in the UK is Alton Towers, which is located in a rural area of Staffordshire and attracts about 2.5 million visitors each year.

Theme parks have become very popular in East Asia. Tokyo Disney opened in 1983, and its latest addition is Tokyo DisneySea. This new development is strategically targeted at an audience to reflect the fact that the Japanese population is ageing. Visitors are conducted through a series of themed lands, which includes a 1900s reconstruction of an American eastern seaboard waterfront complete with steamships. The centrepiece is a 'mysterious' land where an artificial volcano emits steam, and visitors can be taken on a trip to the centre of the Earth.

China is also investing in theme parks, one of the latest to be developed is located in Shunde City in Guangdong Province. The park is themed around the Peanuts cartoon character 'Snoopy' and will cost US$12.1 million to build.

Euro-Disney opened near Paris in 1992 and initially made a loss because the public viewed the attraction as too expensive. Subsequently, remarketed as Disneyland Paris it has since become profitable. The latest addition to the complex is the Walt Disney Studio Park, which opened in 2002 at a cost of US$540 million. Like Tokyo DisneySea, this attraction is targeted at an older age group. The Studio Park, which shows silent movies as well as more recent films such as *Titanic* is expected to boost overall visitor numbers to Disneyland Paris from 12.2 million to 17 million.

The concept of a **holiday-village enclave** has its origins in the holiday camp, which was developed during the 1930s. Holiday centres such as Butlin's still occupy coastal locations, but self-contained village enclaves have recently been constructed in rural areas. Many are set within forests and consist of holiday chalets, shops and restaurants and indoor leisure complexes. They also offer a range of outdoor recreational activities, such as cycling, and are marketed at families. Originally developed by a Dutch company, Center Parcs, holiday-village enclaves have spread from the Netherlands to Belgium, France and Britain. The first Center Parc village to open in the UK in 1987 was located in Sherwood Forest in 1987. Other centres have since been constructed at Eleveden in Suffolk and Longleat in Wiltshire. The Center Parc at Elvedon was destroyed by fire in 2002, but was rebuilt again at a cost of £45 million and reopened in the summer of 2003. A second company called Oasis has opened a holiday village in Penrith in Cumbria.

Summary

1. The tourist gaze has been redirected towards a number of different forms of leisure and recreation in the last 30 years.
2. The media, advertising and travel companies have been instrumental in developing new forms of tourism and recreation.
3. Heritage tourism has grown in response to improvements in transport and education, and promotion by public and private organisations. It has made landscapes, cultures and historic buildings more accessible to a wider audience and created employment, but critics argue that heritage tourism presents a non-authentic, nostalgic view of the past and stereotypes places and cultures.

Questions

1. Why might a location (a) welcome or (b) be reluctant to become the setting for a film or television series?
2. The tourist gaze within the UK has been redirected towards a number of new types of domestic tourist attraction in the last 30 years. With reference to case studies describe some of these developments and explain the factors that have contributed to their success.
3. What do you understand by the term heritage tourism? With reference to case studies explain its benefits and drawbacks.
4. Suggest why heritage tourism has grown significantly in the last 30 years.
5. Research a heritage attraction or theme park. It could be one mentioned in this chapter, or something you have visited, or has a good website. The focus should be geographical, not a guidebook. It should include reasons for general location and specific siting, nature, and number of visitors, funding, costs and nature of attractions, but not in detail. Evaluate the publicity material and discuss benefits of the attraction to the community and environmental impacts.

4 Urban Tourism and Recreation

1 Urban Tourism

a) Characteristics and growth of urban tourism

Many tourists visit urban areas to see attractions such as cathedrals, art galleries and museums, while others go to attend concerts, festivals or sporting events. Some travel to visit friends and relations, or to attend business conferences and trade exhibitions. Often visitors come for more than one reason, for example a business trip may be combined with a visit to an art gallery. Visits are usually short, on average between 2 and 3 days, and not confined to any one season. Urban tourist centres include capital cities such as London and Paris that have internationally famous museums and art galleries, and smaller, older provincial cities, for example Bath and York. Old industrial cities such as Glasgow and Bradford have also become popular urban tourist destinations.

Although cities such as London and Paris have attracted visitors for a long period of time, there has been an increase in **urban tourism** in the last 30 years. A number of factors have contributed to this growth. The traditional seaside resort has declined in popularity while urban areas have been promoted as centres with a rich historic, architectural and cultural heritage (see Photo 4). Higher disposable incomes and more flexible working hours have given people the means and the time to take more short-break holidays. Improvements in transport, notably the Eurostar service to Paris, cheap international and domestic flights to cities such Seville and Glasgow, and improved motorway links have made a short or weekend break more feasible than before. Travel companies have promoted the city break (see Photo 5) and this type of holiday has become a popular competition prize. Cities such as Birmingham and Glasgow have become attractive places for business to hold conferences, trade fairs and exhibitions because they are well served by roads, railways and airports, and have a good stock of high-quality hotel accommodation. **Business tourism** is welcomed by local councils and private enterprises because the customers are usually high spenders.

In depressed, old industrial cities such as Liverpool, urban tourism has created employment and enabled the regeneration of derelict buildings and waterfronts. Tourism is particularly beneficial in these situations because it is a labour-intensive industry; it has low capital costs and produces multiplier effects. Local councils and development agencies have spent money on restoring public buildings, upgrading museums, pedestrianisation, tree planting, street furniture and floodlighting in the hope that private investment would follow in hotel construction, catering and entertainment.

The old harbour area of Baltimore, in the USA, was one of the first urban areas to be regenerated in 1964. The success of this scheme led to many similar projects in the USA. In the UK, industrial cities such as Liverpool and Glasgow went into decline in the 1960s, and the recession in the 1980s left many in need of social and economic regeneration. The case study of Glasgow illustrates how tourism helped to regenerate this city. One of the latest schemes is to re-develop the Melbourne docklands in Australia, at a cost of £2.4 million, into film studios, promenades, cafes, restaurants and shops.

Individual or 'hallmark' events have also helped to promote urban tourism. For example, it is estimated that the Sydney Olympics helped to attract 4.9 million tourists to Australia in 2002 and generated an estimated US$15 billion in export earnings. The surge in visitor numbers began when the Games opened in September and continued to increase towards the end of the year. Barcelona, which held the games in 1992, also witnessed a post-event increase in tourism and has continued to capitalise on this since (see Barcelona case study). Other cities which have hosted the Games, such as Atlanta in the USA, have, however, been less successful in attracting tourism after the event. London is currently bidding to host the Games in 2012 in the hope that it will boost tourism and regenerate the area to the east of the City. Meanwhile, Liverpool has recently been named 'European Capital of Culture' for 2008, a hallmark event, which it is hoped, will boost tourism to the city (see European Capital of Culture case study).

Millennium projects have provided cities with a number of permanent tourist attractions. For example, a new art gallery displaying the works of LS Lowry was built in Salford in the north-west, while London gained Tate Modern. Bath will benefit from the opening of 'Thermae Bath Spa', a project to restore natural hot springs and create an open-air pool, massage rooms, restaurant and shops. The project has cost £23 million and was funded by the Millennium Commission, Bath and North East Somerset Council and the Thermae Development Company.

Photo 4 HMS *Victory* in Portsmouth's maritime conservation area

Photo 5   Venice, a popular city short-break holiday destination

b) Criticisms and future of urban tourism

Urban tourism cannot hope to solve all the problems of the inner city, but it can raise profiles and diversify economies. Critics of urban tourism argue that the jobs it creates are poorly paid, unskilled and only seasonal, and that councils spend money on developing attractions for a middle class while ignoring the needs of the poor. They suggest that urban tourism creates enclaves of wealth in what are otherwise depressed areas giving a misleading impression of city life. They argue that the benefits of tourism are also not set against the cost of increased congestion and wear and tear on the urban fabric. They suggest that tourism causes conflicts between residents and visitors because shops stock gifts and souvenirs rather than essential supplies and hotels occupy land needed for housing.

Improvements in transport and changing political situations have brought new short-break city destinations into being. For example, visitors to Prague have significantly increased since the collapse of Communism in 1989. Hotels have been constructed to meet the increased demand for accommodation and fast food outlets have appeared throughout the tourist area. Outer and ring roads have been improved to reduce city congestion and a new air passenger terminal has been constructed. Further north, the Baltic Republic capitals such as Riga and Tallinn have grown in popularity since the break up of the Soviet Union.

CASE STUDY: GLASGOW

Cotton and tobacco, followed by shipbuilding, locomotives and iron and steel made Glasgow a prosperous city before 1914. The decline of these industries created unemployment and derelict buildings. Glasgow District Council viewed tourism as means of regenerating the city. Early encouragement to achieving this goal came when a benefactor provided money to build a new gallery to house the Burrell Collection, an important collection of paintings and artefacts in Pollok Country Park which opened in 1983. An international airport and good road and rail connections further helped to promote the city. The Scottish Development Agency, the Glasgow District Council and the regional Strathclyde Council helped fund the removal of derelict buildings, the development of new attractions and provided support for business tourism. They contributed to the Scottish Exhibition and Conference Centre opened in 1985 and a Garden Festival held in 1988.

Attractions that have been developed include: the Kelvingrove Art Gallery, the Museum of Transport, an international concert hall, the St Mungo Museum of Art and Religion, a modern art museum, a science museum and an international sports arena. The Scottish Opera, the Scottish Ballet and a Scottish National Orchestra have also been established with help from the City Council. Large shopping complexes have been constructed near St Enoch's and Princes Square, and the City Council has refurbished the physical environment around the Cathedral precinct. The Lighthouse contains an interpretative centre celebrating the art nouveau designer Charles Rennie Mackintosh, and his works can be seen at several venues in the city including the Glasgow School of Art. Glasgow was chosen as European City of Culture in 1990, an award that generated 5580 jobs and £14.3 million in investment. During that year, international arrivals rose by 50%, theatre attendances by 40% and business tourism doubled. In 1999, Glasgow was nominated as the UK City of Architecture, a designation that further helped to boost tourism. Glasgow was the fifth most popular city in the UK visited by overseas visitors in 2001, receiving 400,000 international arrivals. Tourism is the third largest employer in the city after public administration, and banking and finance.

Despite these successes, problems still remain. Once big events finish, visitor numbers quickly fall off and therefore the city must constantly try to reinvent itself if it is to remain competitive. Reliance on business tourism makes it vulnerable when there is a recession. Critics have also argued that the city is too focused on promoting business tourism and developing attractions for the middle-aged and higher income groups at the expense of the less well off.

CASE STUDY: BARCELONA

Barcelona is located in Catalonia in north-east Spain on a coastal plain backed by hills. Trade fairs, notably the Universal Exhibition of 1888, and the International Exhibition of 1929, attracted early visitors, but before the Olympic Games of 1992 Barcelona was still regarded as a commercial and industrial city. The build up and celebration of the Games had important consequences for urban tourism. Hotels, which before 1992 were in short supply, were built or refurbished and the waterfront was physically transformed. Dilapidated warehouses, rail and port installations were converted into promenades, marinas and an Olympic village. Polluted sea water and uncared-for beaches were restored for safe bathing, and public parks and squares were improved. A new terminal at the city airport was built, the railway station was refurbished and new ring roads around the city were constructed. Visitor numbers rose in the year following the Games and since then the Chamber of Commerce and local government has worked hard to maintain this position.

A warm summer climate, a Mediterranean seafrontage and a rich architecture, together with good air, rail and road links to the rest of Spain and Europe, help to explain why Barcelona has grown in popularity as a short, 2–3-day city-break holiday. Visitor numbers rose from 1.7 to 3.3 million between 1990 and 2001. Top attractions include Gaudi's unfinished basilica of the Sagrada Família, the Picasso and Miró museums and the Museum of FC Barcelona. Gaudi's architecture is displayed throughout the city in buildings such as Casa Batlló, Casa Milà, and sculptures in Parc Güell. Other attractions include the medieval Gothic quarter and the 'Eixample', an area of nineteenth century elegant terraces and tree-lined roads. Since the Games more hotels have been built and the Port Vell development has opened on the seafront funded by private and public money. This complex includes a huge aquarium, an IMAX cinema, an international trade centre, restaurants and shops. Strategies to maintain and expand the tourist base include plans to develop the port as a stopping-off-point for cruise ships and promotion as a destination for educational visits.

CASE STUDY: EUROPEAN CAPITAL OF CULTURE 2008

Athens was the first city to take the title 'European City of Culture' in 1985. Since then, competition between rival cities for the award has been intense because of the benefits derived in terms of investment and tourism (see Glasgow case study).

From 2005 the title will be known as the European Capital of Culture and EU countries will take it in turns to nominate a city for 1 year. Ireland has been given the choice for 2005 and has nominated Cork, while Liverpool will take the title in 2008. Liverpool's attractions include The Walker Art Gallery, the Anglican and Roman Catholic Cathedrals, The Liver, Cunard and Port of Liverpool buildings on the waterfront (nominated as a World Heritage Site), The Albert Dock, Tate Liverpool and the recently opened £10 million Film, Art and Creative Technology Centre (see Figure 9 and Photo 6). The city intends to invest £2 million in the next 5 years regenerating the area and improving the physical environment. Money will be spent on public squares, open spaces and street sculptures, and also on supporting art and cultural projects. £750 million has been allocated to

Photo 6   Pier Head, Liverpool

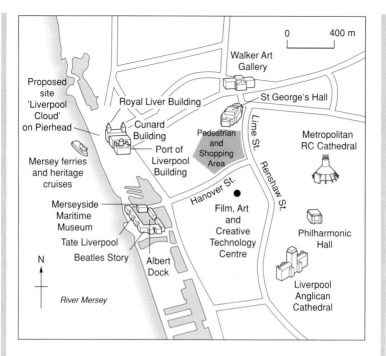

Figure 9 Location of tourist attractions in Liverpool

transforming the city retail centre, and £250 million will be invested in 'Liverpool Cloud', a new visitor attraction on Pier Head outlining the city's history and culture. It is hoped that 14,000 new jobs will be created and 1.7 million visitors will come to Liverpool in 2008.

The importance of the award in promoting urban tourism can be illustrated by examining the impacts on Weimar, a small town in Germany, which won the title in 1999. The town already had many associations with artists, composers and philosophers including Cranach, Nietzsche, Liszt, Goethe and Schiller, but in 1999 visitor numbers soared to 640,000. Some £5 billion was invested in upgrading the urban fabric and improving facilities, which was a large sum for a town with a population of only 62,000. Hotels and the National Goethe Museum were renovated, a conference hall and new museum of modern art were built, and the railway station was restored.

2 Urban Recreation

The Victorians were responsible for the provision of many public parks, squares, walks, recreation grounds, baths and libraries still seen in towns and cities throughout the UK. Local authorities added playing fields and sports pitches to suburban areas during the inter-war years. Since 1945, indoor sports, leisure centres, shopping malls and heritage attractions have been added to the urban fabric, while restaurants, shops and marinas have been common features of waterfront regeneration schemes.

Imposed on this historical development of urban recreation are also spatial patterns. City centres are very accessible locations and therefore command high rents. Major galleries and theatres occupy central locations within cities because they need to be accessible to a wide customer base and the high threshold population is able to meet the high rental cost. In contrast, land on the edges of cities is much cheaper and therefore attractive for the development of large sports centres, golf clubs and out-of-town retail parks.

A number of changes have taken place in urban recreation in recent years. More emphasis on health-related fitness has led to the construction of indoor leisure and sports centres, swimming pools and gymnasia. Many of these complexes have bars and restaurants enabling people to socialise around sport. Shopping as a leisure activity has grown as more indoor malls have been built. Cinemas have been refurbished and many have multi-screens, which have helped to revive film going. Greater varieties of restaurants have opened because people are more affluent and have more varied tastes than formerly. Coffee shops have also multiplied on the high street, both trends reflecting the growing interest in eating out as a recreational activity. Exploring local urban environments on foot has also been popularised by town trails produced by tourist information centres and publications such as *Walking London* by Andrew Duncan.

In contrast, some urban parks and recreation areas have declined in popularity because they have become run down and the public are concerned about issues such as personal safety, vandalism and dog fouling. Some well-maintained parks still, however, remain very popular and perform an important function in providing a green space for the urban dweller. The facilities and management of an urban recreational area are explored in the case study on Richmond Park.

CASE STUDY: MANAGEMENT OF RICHMOND PARK

Richmond Park covers an area of 954.6 hectares in south-west London. The land rises to 56 m above sea level and is covered with oak woodland, acid grassland and bracken. Charles I enclosed the area with a wall creating a royal hunting park in 1637. The public gained access to the park in 1758, although their movements were restricted to roads and paths. After 1851 they were allowed to roam freely. The park was designated a Site of Special Scientific Interest (SSSI) in 1992 to protect its acid grassland and wetland habitat.

Surveys undertaken in the 1990s revealed that 3–4 million people visited the park each year, most arriving by car. About half came at least once a week and 90% lived locally or in the Greater London area. Most people visited the park to walk, although in good weather nearly a third sat or played informal games. Other visitors came to cycle, ride a horse or jog. Most people stayed for between 1 and 2 hours, and many used the refreshment facilities. Many visitors valued the open space and peace and quiet that the park provided, and few wanted the landscape to be more formally managed.

The provision of recreation and conservation within the park is the responsibility of the Royal Parks Authority. Park policy is to maintain the wilderness appeal, which is so valued by the users. The only formal landscapes are two golf courses on the park perimeter and gardens around Pembroke Lodge and within the Isabella Plantation (see Figure 10). Managing carrying capacities and resolving conflicting visitor needs are central to the work of the Park Authority.

The maximum number of users the park can support, or its **physical carrying capacity**, in reality is determined by the number of parking spaces because most visitors arrive by car. The park has 1840 spaces for cars, although untidy parking may reduce this to as little as 1600. Assuming an average length of stay of 2 hours, and the site open for approximately 8 hours a day, then the throughput, or daily capacity, would be 6400. Carrying capacity is, however, a relative rather than an absolute concept and is dependent on how the site is managed. Adding further car parks, or upgrading existing surfaces to accommodate more vehicles, can increase capacity. Alternatively, capacity can be reduced by the closure of car parks. For example, Pen Ponds car park in Richmond is sometimes shut to reduce pressure on the surrounding area. Control of car park space is therefore used as a conservation tool.

The **environmental carrying capacity** of the park, or the maximum number of visitors the site can support without significant damage to its ecosystems, is controlled by a number of factors. These include the number of visitors and the nature of their activities, season of visit, topography, soil, vegetation and geology. For example, people on horseback cause more damage to vegetation than those on foot; wet soils are more susceptible to damage than those that are dry; and some plants are more resistant to trampling than others. Footpath erosion, litter, noise, fumes, removal of rock or plant specimens and disturbance to wildlife are all indications that environmental carrying capacity is reached.

Eroded pathways in the park occur on steep slopes near Pembroke Lodge, near car parks and *en route* to popular attractions such as Pen Ponds. The Park Authority has resisted the creation of formal paths because they would destroy the wilderness appearance and encourage visitors to cluster at certain points. To reduce erosion some paths, such as the Queen's Ride, are regularly reseeded. Cycling is confined to roads and an upgraded gravel path, which runs around the edge of the park. Drainage has been installed along a popular path linking Broomhill car park to the Isabella Plantation.

Despite high visitor numbers, litter does not appear to be a problem, perhaps because visitors value the landscape. The existence of a road, which runs inside the park boundary, is, however, a highly contentious issue. Many drivers use the road as a shorter, more attractive route between A and B and do not stop in the park. Surveys have shown that about 80% of weekday and 60% of weekend drivers use the road in this way. The road is relatively narrow and traffic can quickly build up at park entrances on warm, sunny weekends causing noise and air pollution. Cyclists also at times cause road congestion. A 30 mph speed limit has been imposed on the perimeter road to control vehicle flows. Trees have been used to screen car parks. A further source of noise pollution is aircraft passing overhead *en route* to Heathrow, which is 13 km to the west of the park.

The park is home to 300 Red and 350 Fallow deer. Drivers are warned to be on the look out in case these animals stray onto the perimeter road, but inevitably accidents occur. The public are told not to feed the animals and to keep their distance from the deer, especially when the young are born and during the rutting season, but sometimes these warnings are ignored. The public are also discouraged from straying into areas where skylarks nest on the ground and where gorse and hawthorn are being re-established. The practice of removing dead wood for firewood is also discouraged.

Perceptual carrying capacity, or level of tolerance to crowding, varies spatially across the park. In areas such as the Isabella Plantation, which receives large numbers of visitors in May when the azaleas and rhododendrons are in bloom, tolerance of crowds is quite high. In contrast, in more remote parts of the park, such as White Lodge Plantation, where people seek a wilderness experience crowd tolerance thresholds are much

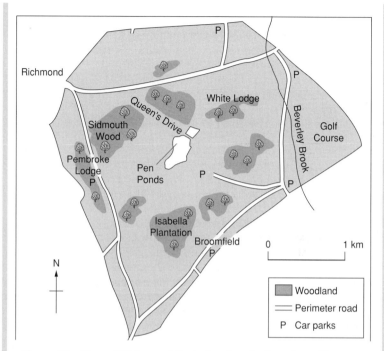

Figure 10   Map of Richmond Park showing the location of features
mentioned in the text above

lower. Average visitor density across the park on Sundays has
been calculated at 5 per hectare which may appear low compared with central London parks but even 1 per hectare can
seem high in a park valued for its wilderness qualities.

In an effort to reduce conflicts, which arise between different
park users, the Royal Parks Authority has adopted a **zoning**
policy. Horse riding is restricted to rides and roads and can only
occur between specified times. Cycling, as already mentioned, is
only allowed on roads and the perimeter gravel track. Ball
games, picnics and bicycles are prohibited in the Isabella
Plantation and dogs must be kept on leads. Smooth, level, gravel
paths within the Isabella Plantation make the area popular with
families with pushchairs and the elderly. The area is also near car
parks, one of which is restricted to disabled driver use only. In
contrast, 'hikers' use less accessible parts of the park where paths
are rough and poorly defined. Managing surfaces and controlling accessibility are therefore both ways by which the Park
Authority is able to control recreational activities that otherwise
might lead to conflicts and damage the environment.

Summary

1. Urban tourism has grown significantly in the last 30 years.
2. City councils have used tourism as a means of regenerating depressed inner-city environments.
3. Critics argue that urban tourism creates poorly paid, unskilled jobs, leads to conflicts between residents and visitors, and increased air and noise pollution.
4. Spatial and temporal patterns can be identified in urban recreation.

Questions

1. Why might a local authority (a) welcome and (b) be reluctant to host a hall-mark event?
2. Suggest why urban tourism has increased significantly in the last 30 years?
3. What evidence would indicate that the capacity of a tourist resource had been exceeded?
4. Comment on the issues that may arise in the management of small area for recreation.
5. Using large-scale maps, local guides and telephone directories, classify and then map the distribution of urban recreational and leisure facilities within a small urban area. Describe the historical development of these facilities and suggest reasons for their spatial distribution.

5 Rural and Wilderness Tourism and Recreation

1 Nature of, and Changes in, Rural Tourism

Many urban dwellers in the UK make frequent day or weekend visits to the countryside, for example car-owning London residents may travel to the North Downs, while people living in Sheffield may go to the Peak District. Another category of visitor stays in urban areas from where they make trips to the countryside, for example a Birmingham resident might decide to make Chester his holiday base and then undertake excursions into Snowdonia National Park. Other visitors stay in the countryside for the duration of their visit, for example by renting a cottage in the Lake District.

Visitors to the countryside engage in different and more diverse recreational activities than they did formerly. 'Going for a drive' and 'picnics' have declined in favour of more active pursuits such as cycling and walking. A survey commissioned by the Ramblers' Association in 2000 revealed that 77% of the British population take at least one walk each month for pleasure. Walking has grown in popularity with the middle classes because people have more sedentary occupations than formerly and there is greater awareness of health-related fitness. The Countryside Alliance and British Heart Foundation have initiated campaigns to raise the profile of walking as a means of keeping healthy. Access to the countryside has improved with the establishment of National Trails (see South Downs Way case study) and the **Right to Roam Campaign**. Councils are now actively marketing areas as walking destinations as a means of creating a sustainable tourism industry. For example, Somerset County Council helped to fund the 80 km Parrett Trail, which opened in 1995. This led to an increase in bed and breakfast accommodation, tearooms and supermarkets along the route. In neighbouring Wiltshire, the 144 km White Horse Trail links the white horses carved in the chalk uplands to Avebury stone circle and the ancient Ridgeway track. An EU-funded promotional pack has been produced to accompany the trail that lists where accommodation and refreshments are available *en route*.

Other types of more active recreational activity that have grown in popularity are mountain biking, off-road driving, orienteering, hang-gliding, cycling and water-based sports. Fashion and media promotion has undoubtedly helped to promote these activities, especially to a young male audience. Greater affluence, health-related fitness and advances in technology, which now permit bikes to be ridden over rough surfaces, and canoes to withstand white water conditions, also explain the popularity of these recreational pursuits. Similar reasons also account for the growth of golf, with the clubhouse acting as a social focus. A number of centres have developed to promote active recreational activities (see case study of High Lodge, Suffolk).

CASE STUDY: THE SOUTH DOWNS WAY

The South Downs Way is a 161 km bridleway, which extends from Winchester to Eastbourne, in the counties of Hampshire, and East and West Sussex. For much of its length, it follows the chalk ridge of the South Downs, an area of outstanding natural beauty (see Photo 7). The South Downs Way is one of 17 National Trails and was established in 1972 by linking existing bridleways and footpaths. It is very popular with walkers, cyclists and those on horseback. Attractions along the trail include Neolithic long barrows, Bronze Age tumuli, Iron Age forts, nature reserves, views across the Weald and the chalk cliffs of Beachy Head.

Surveys conducted during the 1990s found that the majority of users were day-visitors who lived within 48 km of the bridleway. Most travelled by car and their average length of stay was 3 hours. The rest were long-distance walkers who spent between 2 and 7 days on the trail and stayed in bed and breakfast accommodation, or at campsites en route. Visitors had few concerns regarding the management of the bridleway, although there had been some conflicts between walkers and cyclists. There was some evidence of footpath erosion near car parks and on slopes.

Photo 7 Recreational users on the South Downs Way

CASE STUDY: HIGH LODGE, SUFFOLK

Thetford Forest is a large area of coniferous woodland, originally planted by the Forestry Commission for timber, on the Norfolk–Suffolk Border. The woodland became a Forest Park in 1990 and attracts 2 million visitors each year. One of the most popular attractions is High Lodge, a recreation centre, located deep within the forest, which received 175,000 visitors in 2001 (see Photo 8). The attractions include a 9.5 km circular, way-marked cycle track through the forest, signposted walks, a discovery trail with interpretation boards *en route*, an adventure playground and a maze. The lodge contains a restaurant and a shop selling information about the forest. A popular new attraction is an aerial assault course, which consists of a series of rope bridges, swings and zip slides linked together high above the forest floor. Similar high-level trekking courses have been designed in Grizedale Forest in the Lake District and Sherwood Forest in Nottinghamshire.

Photo 8 High Lodge Forest Centre

Other changes in rural tourism and recreation have been the growth of theme parks, holiday villages (see Chapter 3) and farm tourism. Farmers have converted barns and cottages to holiday lets and used meadows for caravan and camping sites. Others have diversified into bed and breakfast accommodation. Farm tourism in Europe is especially well developed in Austria and Italy (agroturismo), while in Wyoming in the USA 'dude ranches' have grown in popularity. In the UK farm tourism has been particularly beneficial to small farmers in the Lake District and upland Wales where incomes from keeping livestock have declined.

A variety of commercial attractions have developed in rural areas including steam railways, working farms, wildlife parks and rural museums, many supported with craft shops and restaurants. A typical example is Park Farm, near Snettisham in west Norfolk (see Figure 11), which has attractions including a farm trail, a deer safari, pony rides, a craft shop, a tearoom, a picnic area and an adventure playground, as well as opportunities for children to feed and touch domestic animals. Such attractions provide parents with safe environments where children can play and a day out for the family.

A further development has been the growth of second-home ownership in rural areas. This practice is widespread in Europe, although level of ownership is higher in some countries than in others. Frequency of use also varies, with some people using a second home

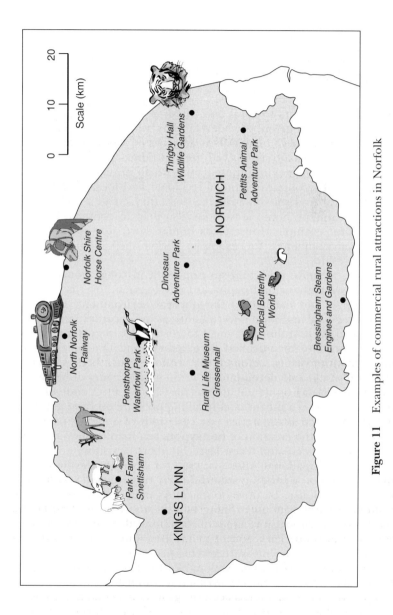

Figure 11 Examples of commercial rural attractions in Norfolk

on a regular basis to compliment a city flat while others make only occasional visits to a country cottage in the summer season. Some people also own second homes in countries other than their normal place of residence, for example UK residents may own another home in Brittany in France.

2 Management of Rural and Wilderness Tourism

The influx of visitors to rural and wilderness areas creates benefits and problems. Tourism provides farmers with another means of making a living. Multiplier effects are created when local growers supply commercial attractions with fresh produce. Rural tourism can lead to improvements in infrastructure, which in turn benefit other sectors of the economy. National, regional and local governments support rural tourism initiatives because they create earnings and local employment. Some attractions also help to conserve national heritage, for example Housesteads Roman Fort, Northumbria, and natural environments, for example Wicken Fen Nature reserve, Cambridgeshire.

On the negative side, tourism can result in congestion on rural roads, unsympathetic and haphazard development, overuse of local water supply and damage to the physical environment. For these reasons regional and local governments, as well as other bodies such as National Park Authorities, need to ensure that regulations are in place to prevent uncontrolled development and that new projects integrate with existing sectors of the economy. Such bodies can also advise on sustainable development, devise tourist codes of behaviour and carry out environmental audits (see Chapter 6). A code of practice is also used as a means of controlling pressure on remote wilderness regions such as Antarctica (see case study of Antarctica)

Zoning and the creation of **honeypots** can help to alleviate some of the problems created when large numbers of visitors, some with conflicting recreational interests, visit rural and wilderness areas. Zoning involves separating out different recreational activities or levels of visitor use, by time or space. As previously discussed, the Royal Parks Authority uses zoning as a means of preventing undue pressure on more sensitive areas of Richmond Park (see Chapter 4). Dartmoor National Park Authority has identified areas of high and low tolerance to visitor pressure and has planned facilities accordingly (see case study of Dartmoor National Park). The Great Barrier Reef in Australia, which is a huge tourist attraction, has been zoned into a number of areas. These include: preservation zones where no use is allowed, scientific research zones where research is permitted under strict controls, marine national park zones where scientific, educational, recreation use is allowed but removal of resources is prohibited, and general use zones where commercial and recreational fishing can take place provided it is undertaken sustainably.

One approach to zoning is to distinguish recreational activities that can use the same space without conflict from those which are incompatible. Different water-based recreational activities often conflict with each other. One solution is to zone the water resource so that different water sports occupy different areas, for example at the Cotswold Water Park, Gloucestershire. Another solution is for motor cruisers and sailing boats to use the waterway at different times of the day.

Honeypots are sites designed to deliberately draw large numbers of visitors so that pressure on the surrounding environment and local communities is reduced. For example, **Country Parks**, which were established in 1968 and now total 350, were originally designed to act as focal point for countryside visits. They were also set up to relieve pressure on sensitive locations within National Parks. This policy failed and honeypots have now been developed within national parks. One example is Danby National Park Centre in Cleveland, which draws visitors from across the North York Moors.

Another approach to visitor management is to disperse tourists so that their impact is more evenly spread across the site. One way of achieving this is to design scenic drives which convey tourists to visit less frequented areas. A further approach is to restrict visitor numbers through pricing or exclusions, for example US National Parks charge an entrance fee and some entrances are closed during the winter (see Yellowstone case study).

3 The Role of National Parks in the UK

National Parks were created by the 1949 National Parks and Access to the Countryside Act, and the first park to open was the Peak District in 1951. Most National Parks are located in the north and west of Britain and cover large areas of high landscape value. Much of the land is privately owned and is used for a variety of purposes including water supply, forestry, farming and mining. The body responsible for park administration and planning is the National Park Authority, an organisation largely funded by government. Its aims of designation, as defined by the 1995 Environment Act, are to conserve and enhance the natural and built environment, to provide access for recreation, and to support and protect the economic and the social well being of local communities.

National Parks have increased in popularity and now receive about 100 million visitors per year. This has resulted in road congestion on popular routes such as the southern approaches to the Lake District and erosion of footpaths such as on the Pennine Way in the Peak District. Suggested solutions to these problems include closing car parks and putting on buses to popular honeypot sites instead, spatial and temporal zoning, and charging entrance fees. One proposal currently under consideration is to impose a £3 charge on vehicles travelling up the road to the Ladybower Reservoir in the Peak District, a popular tourist venue.

CASE STUDY: DARTMOOR NATIONAL PARK

Dartmoor National Park covers an area of 954 km² in Devon. A granite plateau rising to 335 m topped with tors underlies more than half the land surface. Blanket bog occurs in the western part of the upland, grass heathland is found in the east, while the edges of the moorland are fringed with oak woodland.

Tourism dates from the early 1900s when visitors stayed in hotels in such places as Chagford, on the eastern edge of the moor. Coach trips were organised to popular spots such as Dartmeet in the 1930s, and the area became a National Park in 1951. Within the park the land is privately owned but the public are allowed to roam across unenclosed, open moorland. The public have access to 47.5% of the Park and there are also 660 km of rights of way. Land in the park is used for a variety of uses including rough grazing, reservoirs, military training, mining and forestry.

The membership of the Dartmoor National Park Authority is drawn from Devon County Council, government appointees and people with special knowledge or expertise of the area. The Authority manages four information centres and an education service, and provides rangers who patrol the area and offer guided walks. Central government provides three-quarters of funding for these facilities and the rest comes from county, borough and district councils.

Visitor surveys suggest that almost 90% of those coming to Dartmoor travel by car. Most people come to walk, or visit attractions such as Castle Drogo, Buckfast Abbey and Lydford Gorge (see Figure 12). On average, about half of the visitors live locally and the rest visit the park while staying elsewhere in the region. Visitor numbers have increased from approximately 8 to 11 million between 1980 and 2000. Over this period, the park has become more popular for active recreational activities such as cycling, hiking, horse-riding and cycling. Tourism was badly affected by the 2001 outbreak of foot-and-mouth disease.

Rising visitor numbers, high concentrations of tourists during the peak July–August season and a greater emphasis on active recreational pursuits have placed an increasing pressure on the environment. Riverbanks and footpaths have been eroded and some stones at archaeological sites have been disturbed. In response to visitor pressure the Park Authority has established a number of guidelines and put a number of initiatives into place. A 40 mph speed limit has been imposed on roads within the park, and the Sunday bus service has been improved. Heavy traffic is encouraged to go around, rather than attempt to cross, the moor. Camping on the roadside or in car parks is prohibited. Parking is allowed in designated areas but not on verges. Hikers are encouraged to walk on hard surfaces when the ground is wet to reduce erosion, and to use gates rather than climb over stone walls. Visitors are asked to consider using public transport rather than coming by car, and to avoid starting fires which can quickly spread when the vegetation is dry. Cycling is confined to bridleways and roads to avoid erosion of common land.

An EU-funded 'Moor Care Programme' was started in 1996 with the aim of repairing and upgrading sites and paths, and educating users about sustainable practices. One area to benefit from the scheme is Hay Tor Rocks on the eastern side of the moor. The rock outcrop is popular with climbers and there are the remains of a nineteenth century granite horse-drawn tramway nearby. Erosion has occurred on paths up to, and around, the base of the viewpoint. Cars parking on the roadside damage the moor and horse-riders have also contributed to erosion. Restoration has involved levelling, turfing and reseeding paths and fencing the area near the tramway. In an effort to spread visitor impacts a new path up to the tors has been created and signposting of routes has been improved.

The Park Authority have introduced a zoning policy across the moor in an effort to reduce pressure on sensitive sites. Visitor facilities have been built at sites most able to sustain high visitor numbers, while these have been kept to a minimum at more sensitive areas. The timing, scale and location of organised events such as the Ten Tor's Challenge is carefully managed to ensure that pressure on the environment is minimised.

Plans for the future include improving off-road cycling in a sustainable manner and making woodland more accessible so that this provides horse-riders and cyclists with a wet-weather alternative. Other objectives are to consider ways of developing recreational opportunities on the edges of the park and to integrate public and private sector bodies engaged in marketing Dartmoor through a forum called 'Dartmoor Partnership'. The Park Authority is also keen to promote initiatives that help tourists learn more about the moor and the communities within it and to encourage those marketing tourism to conduct environmental audits of their businesses.

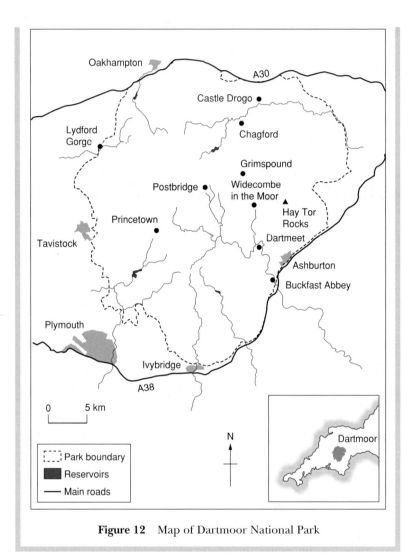

Figure 12 Map of Dartmoor National Park

CASE STUDY: MANAGING YELLOWSTONE NATIONAL PARK, USA

Yellowstone Park is located mainly in the state of Wyoming and is a UNESCO World Heritage Site and a World Biosphere Reserve. Its many physical attractions include a large number of active geysers, hot springs, mud pots and steam vents, dramatic waterfalls and canyons. Located within coniferous forest and open parkland, Yellowstone also has abundant wildlife including bison, elk, bear, moose, wolf and coyote, together with trout-filled rivers and lakes fringed with swans and pelicans. The National Park Service has the dual role of preserving the natural and cultural resources and also making these accessible to the public. These goals are often incompatible and difficult choices have to be made.

The park was established by Congress in 1872 and initially inaccessibility kept visitor numbers low. Wealthy Americans alighted at rail stations just beyond the Park boundary and were then taken by stagecoach to luxury hotels within Yellowstone. The first car entered Yellowstone in 1915 and by 1940 this became the main mode of transport to, and within, the park. Cars improved accessibility and broadened the social class of visitor, which led to mass tourism. Visitor numbers rose from 18,000 to about 1 million between 1908 and 1948, and reached 2.25 million by 2001.

Inevitably large numbers of visitors, many arriving during the short summer season, have created problems. At popular attractions such as the Old Faithful geyser several thousand people gather to see the geyser erupt. Car parks become very crowded creating air pollution and large crowds can detract from the aesthetic attraction of the site. Tourists are told to stay on the boardwalks and not to throw objects into the geysers in the hope that these will improve the blast but nevertheless this has occurred damaging the vents.

Most visitors view the park from the roadside rather than hiking into the backcountry. Roads, which were originally built for stagecoaches not cars, are now in a poor state of repair, although a rebuilding programme is underway. A 45 mph speed limit is strictly enforced to prevent accidents between animals and cars but inevitably these sometimes occur. Traffic jams quickly develop when an elk or a bear is spotted near the roadside. Visitors who have got too close to a buffalo, which are numerous in the park, have been gored. Feeding bears, which was once a common practice, is now strictly forbidden because it leads to habituation, a process by which bears become reliant on humans for food. When this is denied, the bears become aggressive and cause damage to property in order to gain access to supplies. To reduce habituation, waste dumps within Yellowstone have been removed, bear-proof bins have been installed and troublesome bears have been removed to more remote areas.

In a further effort to reduce conflicts between bears and humans, the campground at Fishing Village was closed because it was located in prime grizzly bear habitat. The bear's favourite food was the cutthroat trout, which was common in rivers nearby.

Some buildings were subsequently relocated to Grant Village but this proved an equally contentious site because it was also located in grizzly bear habitat. Other concerns were that that poorly treated human sewage from the village might pollute Yellowstone Lake and that a planned marina would alter wave currents leading to beach erosion. To reduce boat engine noise and beach erosion the Park Service has banned powerboats from some parts of the Lake and in other areas boat speed limits are enforced. Large boats are prohibited from approaching lake islands because they disturb nesting sites.

The introduction of snow-coaches and more recently snow-mobiles has dramatically increased winter visitation to about 140,000 annually. The noise from the snowmobiles together with the blue smoke emitted from their fuel-inefficient engines disturb bison and elk and create air pollution, especially on roads near the park entrances. Efforts by the Park Service to ban snow-mobiles from the park have met with strong opposition from the business community in West Yellowstone who profit from this recreational activity. Proposals to limit or phase out snow-mobiles, or to restrict use to cleaner, less noisy machines are currently under discussion.

Without adequate funding, there are no easy solutions to the growing visitor pressure on Yellowstone. Options include limiting the number of visitors, or vehicles, that enter the Park each day, closing campgrounds and promoting the backcountry. If cars were banned and replaced by buses or a monorail, which operated from the gateway towns, the service might need to be subsidised and a new road network constructed. More visitor provision would also be needed at the gateway towns and more visitor pressure would occur at popular sites. Moreover, people would have insufficient time to the view geysers, many of which are more impressive than Old Faithful when they erupt, or experience the sounds and smells of the forest, or see animals in their natural habitats at dawn or dusk. Removing all overnight camping in the park would reduce visitor flexibility, overburden roads and require more visitor facilities to be built at gateway towns. Restricting the number of cars entering the park to a first come, first serve basis might be a way forward, but is likely to be met with opposition from those who perceive tourism as an important and growing source of revenue to Wyoming.

Promoting the backcountry would be unpopular with visitors, the majority of whom prefer to see the sites of Yellowstone by taking short walks along paved paths from their cars. Visitors currently using the backcountry would also be against such a proposal because the wilderness experience they seek would be lost.

CASE STUDY: ANTARCTICA

Antarctica is becoming an increasingly popular tourist destination. Over 95% of visitors view the continent from the comfort of a cruise ship, and make occasional trips onshore for a few hours to see penguin rookeries, scientific research stations, former whaling stations and early explorers' huts.

Seaborne tourism began in the 1957–8 summer season, but visitor numbers have recently risen sharply from 6000 in the 1991–2 season to 10,590 in 1997–8. The number of sites visited has also significantly increased with popular localities, such as Half Moon Island off the Antarctic Peninsula, receiving tourists every 2–3 days during the short summer season. The cruise ships carry between 40 and 450 passengers, and depart mainly from South American ports to the South Shetland Islands, the South Orkney Islands, the Antarctic Peninsula and South Georgia (see Figure 13). These localities have a mild summer climate, abundant wildlife and are free from pack ice, which together with their close proximity to South America makes them popular tourist destinations. Other ships depart from Australia and New Zealand and travel via the sub-Antarctic Macquarie Islands, to Cape Adare, McMurdo Sound and the Commonweath Bay Sectors of the continent. The journey takes 10 days; much longer than the 48 hours needed to cross the Drake Passage from South America to the Antarctic Peninsula, which explains why fewer boats undertake this voyage.

The remaining 5% of tourists are airborne travellers. Some fly from South America to a 'guest house' on King George Island off the Antarctic Peninsula. Flying avoids the necessity of crossing the Drake Passage, which can sometimes be rough. Once on the island visitors are taken to see glaciers, penguin rookeries, whaling stations and elephant seal colonies. Sightseeing flights over Antarctica were curtailed in 1959 after a New Zealand airliner crashed into Mount Erebus. Today Qantas Airlines offers sightseeing flights over the continent using Boeing 747s, which fly well above the Antarctica's mountain peaks. Flights depart from Sydney and last about 12 hours, of which 3–4 hours is spent above the continent. One company called Adventure Network International offers private expeditions in Antarctica. From the company's base camp at Patriot Hills at the foot of the Ellsworth Mountains participants can climb Mt. Vinson, Antarctica's highest peak, or travel by small plane to the South Pole.

Some benefits have accrued from the development of Antarctic tourism. For example, discarded rubbish at research stations on King George Island was cleared up once the tourists started to arrive. Visits to research stations have helped promote an understanding of their work. On-board lectures have raised public awareness of sensitive issues such as whaling. Tourism has also helped to boost the economies of staging posts such as Ushuaia in South America, a settlement where there are few other job opportunities.

Research into the impacts of tourism conducted in the 1990s largely concluded that tour operators and visitors were acting responsibly and that, thus far, the natural environment and wildlife had not been adversely affected. Nevertheless, there are concerns that if visitor numbers continue to increase some negative impacts will occur. Moss and lichen grow very slowly in the harsh climate and are therefore highly sensitive to trampling. The short, summer tourist season coincides with the breeding period of many birds, which raises concerns that visitors may disturb penguin rookeries. Arctic terns appear to be unsettled by human presence and tourists have already damaged exploration huts.

Currently, management of Antarctica is based on guidelines. No one country can impose regulations because the Antarctic Treaty put aside the issue of territorial claims. Many cruise ship companies are members of the International Association of Tour Operators, an organisation that has established a code of practice on matters such as the disposal of waste. Experienced staff operate the zodiacs, which carry passengers to the shore. The recommended maximum number of visitors ashore at any one time is 100 and their movements are closely supervised. Before going ashore tourists are briefed on the requirements of the Antarctic Treaty and the Environmental Protocol Code. Tourists are told not to encroach within 5 m of a penguin and 15 m of a fur seal and not to feed the wildlife. They are reminded not to trample vegetation or remove 'souvenirs' such as feathers, whalebone, rocks and plants, and not to drop litter.

As more ships visit Antarctica there are calls for a scheme of licensing to be introduced. This would help to ensure that waste is properly disposed of, and that only ships with ice-strengthened hulls operate in Antarctic waters thus reducing the risk that they might go aground and cause an oil spill. One large operator has recommended that ships carry a maximum of 140 passengers to limit the impact on the environment. Smaller numbers would also make it easier to evacuate from the land should the weather quickly deteriorate.

So far proposals to establish more land-based accommodation for tourists have been resisted. An Australian company in the late 1980s wanted to build facilities near the Davis Base but the Australian government and conservation groups opposed this because of concerns about the disposal of waste, and food and water supply. It has been suggested that Antarctica should become a World Park but this raises questions as to who would maintain and fund it. So far the remoteness and harsh environment of Antarctica has helped to limit tourist development and where this has occurred operators and visitors have acted responsibility. There are, however, concerns that if visitor numbers continue to grow and/or guidelines are ignored, then environmental damage will follow.

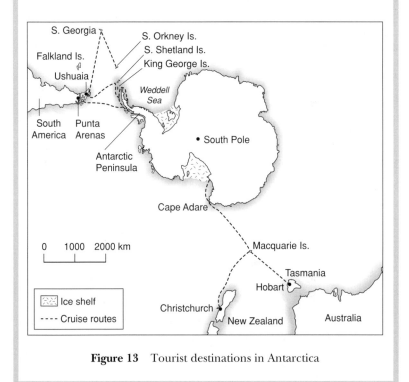

Figure 13   Tourist destinations in Antarctica

Summary

1. Public access to the countryside has increased in the last 50 years.
2. More active forms of recreation such as walking, cycling and canoeing have replaced activities such as going for a drive and picnics in rural areas.
3. Rural tourism has brought benefits and created problems in the countryside.
4. National Parks perform a dual role in providing recreation for users and conserving natural, historic and cultural resources.

5. Approaches to managing large numbers of visitors in rural and wilderness areas include zoning, creating honeypots, dispersing visitors across the attraction, and imposing quotas and entrance fees.

Questions

1. Suggest three reasons why walking has become a popular recreational activity in the UK.
2. What are the differences between a National Park and a Country Park?
3. Using examples, describe and explain why rural tourism has grown in recent years.
4. What factors underlie the changing demand for recreational resources in National Parks and similar areas?
5. Examine the pressures imposed on the capacity of areas such as National Parks, by their intensive use for leisure, recreation and tourism.

6 Economic, Socio-cultural and Environmental Impacts of Tourism and Recreation

1 Economic Impacts

a) Positive economic impacts

Tourism can bring a number of economic benefits to a host country. International tourism generates foreign currency earnings that can improve a country's balance of payments. An indication of how well tourism contributes to gains and losses in foreign currency is the **travel account**, which is calculated as the difference between tourist receipts of a host country and the tourist expenditure of the residents when they travel abroad. Countries such as The Gambia have a positive surplus because it is a popular tourist destination, but most Gambians are too poor to travel abroad as tourists themselves. In contrast, Germany has a negative account because it is not a particularly popular tourist destination and therefore tourist receipts are low, but many of its nationals are affluent and travel widely. It should, however, be noted that More Economically Developed Countries (MEDCs) with negative travel accounts often also control multinational travel companies and airlines, and therefore profits from international tourist earnings can convert **balance of payment** deficits into surpluses.

Money generated from foreign earnings, park entrance fees, as well as accommodation, restaurant, airport, sales and employee income tax revenues can aid economic development. Money can be used to improve transport leading to more trade. Tourist revenues can be spent on sewerage systems, water treatment plants and better housing. Income can contribute to the restoration of historic and cultural monuments. Foreign investment can also often act as a catalyst for growth, particularly in developing countries that lack funds for large-scale capital projects such as hotel construction.

In some countries tourism can make a significant contribution to **GDP**, for example The Bahamas and The Maldives. Tourism creates employment, for example in hotels and transport. It also has **multiplier effects** in that some of the income generated from the hotel bill allows the proprietor to buy food from a local producer, who in turn can buy material goods thus creating more employment. Initially, this type of linkage may be slow to develop because newly established hotels in developing countries often depend on imported goods. As studies in the Caribbean have shown, however, over time links between hoteliers and local suppliers develop which in turn leads to further developments in wholesaling and food processing and less reliance on imported food. The multiplier effect is often expressed as a ratio, for example a ratio of 1 to 1.30 means that for every $1 spend directly on tourism, a further 30 cents is created indirectly in the regional or local economy.

Tourism helps to diversity economies, thereby reducing the reliance on a few primary industries, which might be affected by changes in world demand or competition from cheaper sources of supply. In MEDCs diversifying into tourism has offset falling farm revenues in rural areas. In Less Economically Developed Countries (LEDCs) rural tourism can help to stem the drift of the young, fit and able towards cities in search of work. In MEDCs the development of urban tourism offers a means of regenerating depressed inner-city areas. In these ways tourism can help to redistribute wealth between more and less affluent parts of a country. Tourism can also encourage the transfer of wealth from relatively rich to poorer countries. This occurs, for example, when tourists from relatively affluent northern European countries, for example Germany, decide to holiday in poorer Mediterranean areas such as Greece.

b) Negative economic impacts

Tourism can produce a number of negative economic impacts on a host country. Employment is often seasonal and hotels during the low season close, or are only partially filled, which means capital assets are underused. During the peak season local people may choose to work in tourism rather than farming, which can result in labour shortages particularly at harvest time. Many jobs created by tourism are low paid and unskilled. Host governments, especially in LEDCs, are often reluctant to press international companies for higher wages because they need the foreign exchange earnings. They also recognise that in a very competitive market the foreign tour operator could easily switch to another destination if wage demands rise.

Income generated from tourism is often transferred overseas, rather than benefiting host countries. For example, it has been estimated that in the Cook Islands in the Pacific only 17% of the income derived from tourism benefits local people. Loss of revenue or 'leakage' is particularly associated with all-inclusive package holidays and tourist enclaves. Money is lost when foreign-owned airlines and hotels send the profits they have made in the host country back to their company headquarters. Foreign hotel ownership, and consequent leakage, is common in LEDCs such as The Gambia (see case study), but also occurs in MEDCs such as Spain and Greece. Leakage also occurs because much of the food used in the hotels is imported. Hoteliers are often reluctant to buy from local suppliers because tourists have conservative tastes; the food is poor quality or has been produced under unhygienic conditions. For their part, local suppliers find it difficult to produce high-quality food without adequate storage facilities. They also lack contacts with hoteliers with whom they can conduct business. A further cause of leakage occurs when senior, foreign, hotel staff transfer their earnings to MEDCs. Money is also lost when host governments have to pay high interest charges on money borrowed from overseas institutions to develop tourism.

Over-reliance on foreign investment and expertise in tourism can discourage host countries from developing other sectors of their economy. Reliance on tourism can also be risky because natural hazards, unseasonable weather, political instability and unfavourable exchange rates make the industry volatile to changes in supply and demand.

Tourism can concentrate, rather than disperse, wealth within a country. For example, the development of tourism in The Gambia has been focused on the coast, while the rest of the country which has few tourist attractions has remained undeveloped. In the UK, international tourists tend to visit London, Stratford and York, while remote rural areas remain unvisited. At popular holiday destinations, competition for space can lead to inflated land values and higher prices for goods in the shops.

The tourist industry can also produce some less direct, negative economic impacts. For example, local businesses incur extra costs as the result of tourist-induced traffic congestion in cities. Local authorities have to find extra money to tackle tourist-related increased crime and pollution.

CASE STUDY: THE GAMBIA

The Gambia is a small country in West Africa surrounded by Senegal (see Figure 14). The River Gambia flows throughout its length, fringed by mangrove swamps in its lower reaches. Much of the land is low lying and covered with savanna woodland and grassland. In 2001 the population totalled 1,400,000 and was growing by 3.1% per annum. The Gambia is a very poor country as illustrated by a per capita GDP of only US$1000, an average life expectancy of 54 years and a literacy rate of 48%. It has no mineral resources and 75% of the labour force is engaged in agriculture. Its main exports are groundnuts and groundnut products, fish, cotton lint and palm kernel.

The Gambia has become a popular, winter holiday destination for North Europeans. Located within 6 hours' flying time of London, it has a warm, dry winter climate, palm-fringed beaches and a river environment has been described as an ornithologists' paradise. It has a cultural history that is linked to the slave trade, an episode described in Arthur Hailey's popular novel 'Roots'. Until the mid-1990s The Gambia was regarded as one of the most politically stable countries in Africa. It gained independence from Britain in 1965 and is English-speaking.

A Swedish tour company was the first to bring 300 tourists to The Gambia in 1966. By 1971 visitor numbers had increased to about 8000 and by 1993 this figure had risen to 90,000. About 150,000 tourists visited The Gambia in 2000, about half of which came from the UK, the rest originating from Sweden, Germany and France. Most came on all-inclusive, 1–2-week, package holidays.

The Gambian government has been keen to diversify into tourism because its economy is heavily dependant on groundnuts, a crop that has fallen in value on the world market and is susceptible to disease and drought. Tourism also generates much-needed foreign earnings, currently estimated at about US$25 million. Airport, income, bed and sales tax all provide revenue for the government. Many of the holidays on offer are, however, an all-inclusive package, which means that locals derive little benefit from tourist spending. Only about 30% of tourists take organised tours while staying in The Gambia, and because they have already paid for meals within the hotels, local restaurants are underused and very limited in number. Tourists do not use local taxis because they have already paid for excursions, and are reluctant to use local guides because they have been warned about young men, or 'beach boys', who offer a range of services including prostitution. Moreover, the type of tourist visiting The Gambia tends at be on a relatively low income by European standards and is therefore not a high spender. Furthermore, a majority of the tour operators, hotels and airlines are foreign owned and therefore profits go overseas. Tourism contributes to about 12% of the GNP. About 7000 Gambians are directly or indirectly employed in tourism, although many jobs, for example hotel waiters, are low paid and unskilled. Higher paid managerial jobs are given to expatriates who are regarded as having the education and experience to run hotels. High leakage occurs because manager's salaries are transferred and spent in MEDCs. About half of the local people employed in tourism are laid off at the end of the season. The groundnut harvest coincides with the start of tourist season, which causes some labour shortages in agriculture at this time. Tourism creates work for taxi drivers, guides, and those producing and selling souvenirs. Fishermen also do well out of tourism and farmers supply some local produce, for example watermelons, but about 60% of the food and 40% of the drink is imported. Lack of storage faculties, made worse by the hot climate and unreliable electricity supply, together with the problems of producing food for European tastes, imposes limits on the amount of local food that farmers can produce for the tourist market.

Overseas aid to support tourism development has been used in the coastal area to build roads linking hotels, the airport and urban area. Local people, however, received no compensation for loss of their homes when road construction took place. The government has also had to pay high interest charges on the money they borrowed.

Tourism helped to establish Abuko National Park, a reserve that protects savanna woodland from deforestation. In other respects, however, tourism has resulted in a degradation of the environment. Forty years ago the Kololi coastal area consisted of huts where extended families caught fish and grew cassava, sweet potatoes, maize, rice, and tomatoes and mangoes. When the roads and hotels were built locals lost their livelihoods. They could not obtain jobs in the hotels because these went to those who attended the hotel training school. Beach access is now denied because of hotel development, and tourist facilities make demands on local water supplies. Illegal extraction of beach sand for hotel construction may have contributed to increased coastal erosion.

There is some evidence that tourism has contributed to an increase in crime, drug abuse and prostitution. Local youths patrol the beaches offering tourists a range of services including sex. As a consequence many young people do not attend school and their education suffers. Tourists often dress inappropriately in a country where 80% are Moslem. Native dances, orchestrated for visitors, are performed in hotels. Traditional woodcraft skills have declined as more souvenirs have become mass-produced.

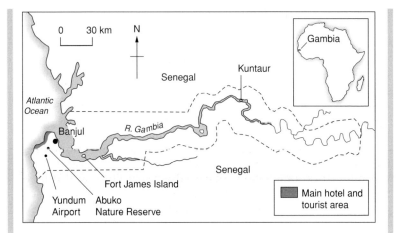

Figure 14 Map of The Gambia and the main tourist area

A number of suggestions have been advanced to help the Gambians develop tourism. Training locals in hotel management and guiding might help them secure better jobs in the tourist industry. Improvement in storage facilities, electricity supplies and modern farming methods would improve the quality of local produce, which could then be sold to hotels. Better quality souvenirs would provide more profit. Diversifying into wildlife tourism, especially ornithology, would help to spread the benefits of tourism more widely.

Sustainable tourism has been attempted in Tumari Tenda, a village in a forest clearing on a tributary of the river Gambia. The 300 inhabitants originally won a prize for forestry conservation, but then moved into tourism in 1993. Visitors stay in cabins, engage in activities such as fishing and cooking, and are taken on guided tours of the local area. The problem, however, is that the scheme does not generate much income and the numbers involved are small.

Poverty and debt underlie the problems of developing tourism in The Gambia.

The government has limited resources to invest in tourism, in a country where money is also desperately needed for education and health care. The Gambia has only one airport, most roads are unpaved and there is a heavy reliance on ferries rather than bridges to cross rivers, which prevents tourism from spreading into the interior. Foreign companies are reluctant to invest in tourism because potential visitors might decide to holiday in more fashionable, or more competitive, sun–sea locations.

2 Socio-cultural Impacts

a) Factors influencing the level of socio-cultural impact

The socio-cultural impacts of tourism are often difficult to quantify and as a consequence have received less attention than economic or environmental effects. Impacts are also difficult to isolate from the social changes brought about by the modernising influences of radio, television and improved communications. The perception that tourism destroys local societies has tended to result in an over-emphasis on negative impacts, whereas in some cases there have also been socio-cultural benefits.

A number of factors contribute to the degree of tourist impact on host communities. In general, large numbers of tourists have more impact than individuals. Impacts are also more pronounced where there is a large difference between the host and tourist cultures. Providing, however, there are opportunities to exchange ideas, cultures can become more similar over time (**acculturation theory**). Large numbers of tourists visiting small communities, such as villages, or islands are more difficult to assimilate and cause greater impacts than high visitor numbers to cities. Enclave resorts, or self-contained hotel complexes, tend to have a minimal impact on host communities, for example Luperon Beach Resort, The Dominican Republic. Different host sectors are likely to react differently to tourism; business and government may welcome tourists but local residents might be negative. Tourists themselves may be drawn from different cultural backgrounds and therefore their impacts will vary depending on the composition of the group. Attitudes to tourists may change over time. Initially, small groups of 'explorer type' tourists may be welcomed because they take an active interest in the host society and they bring economic prosperity to some sectors of the community (see the Butler model in Chapter 1). Over time, however, as numbers increase and more negative impacts emerge, attitudes towards tourists change from slight annoyance through to open hostility, This concept, known as Doxey's index of irritation, has, however, been criticised because it assumes that negative impacts continue to grow whereas successful management can help to reduce tourist pressures. Moreover, greater local community involvement can also help to alleviate negative impacts.

b) Positive socio-cultural impacts

Positive socio-cultural impacts include the notion that tourists acquire a better knowledge, understanding and, sometimes, empathy with host societies and cultures. Another benefit, known as the positive **demonstration effect**, occurs when host societies observe and, in time, adopt higher moral codes of behaviour and/or more productive work patterns as a result of contacts with tourists who hold such values and attitudes.

Tourist expenditure on traditional crafts, and interest in local customs and rituals, may ensure that these survive or are revitalised. In Bali, for example, the development of tourism and the preservation of the culture have been self-reinforcing. Tourism also encourages social empowerment, especially for women. The money local people earn from tourism and the new language skills they acquire give them a greater freedom to choose where they live and their marriage partners.

c) Negative socio-cultural impacts

Tourists who behave badly, or dress inappropriately, create a negative demonstration effect. Young people in host communities are particularly susceptible to outside influences and quickly adopt foreign values, dress codes and lifestyles, some of which may conflict with the more traditional views held by the older generation, for example in The Gambia. More generally, where tourists and hosts are drawn from very different backgrounds then clashes of culture are likely to occur resulting in misunderstandings about customs, religious observances and etiquette. Resentment and envy of tourist wealth can lead locals to demand more luxury and imported goods, which drain the local economy. Young people in search of wealth move from rural to urban areas leaving behind an ageing, dependent community. Rural to urban migration can result in higher rates of separation and divorce (see the case study on Nepal).

Local people may acquire a taste for high-calorie, Western foods, which lead to a rise in heart disease. Smoking and consumption of alcohol may also increase, causing higher rates of lung cancer and liver failure.

The image presented by tourists while on holiday often differs from the pattern of behaviour and consumption patterns exhibited at home. For example, visitors often spend more on luxuries, food and alcohol, and they dress very casually. Unfortunately, the set of behaviour patterns that emerges presents a misleading image to the host community.

It is claimed that tourism promotes antisocial activities such as gambling, prostitution and crime. These activities appear to be increasing in holiday destinations, such as The Gambia, Kenya and the Caribbean. In Thailand, for example, there is evidence that fathers in rural hill villages sell their daughters to sex tourism in order to settle debts owned to moneylenders. Others argue, however, that tourism is not responsible for the introduction of antisocial activities; rather, that it creates conditions that allow existing practices to flourish. For example, differences in wealth between tourists and host communities may lead to a rise in robbery and muggings, and visitors accidentally straying into unsafe urban areas may find themselves victims of assaults.

Tourism can lead to a debasement and commercialisation of indigenous culture. Ritual dances are 'commodified' or packaged in such a way to make them easy to understand and replicate for a tourist audience. Such staged events are often held in hotels, rather than their intended original locations. In time performers lose their understanding of traditional cultural practices and knowledge of local languages and the 'pseudo event' becomes accepted as the norm. Similarly, the mass production of souvenirs results in a decline in craft skills and eventually contrived art is perceived as authentic.

Large numbers of tourists can physically overwhelm small village communities and appear intrusive in cathedrals and mosques. Many tourists perceive churches as visitor attractions, not places of worship. They fail to respect religious ceremonies, which might be taking place, and they disturb those who come to pray.

The construction of tourist facilities such as hotels and golf courses can physically lead to a displacement of local people, destroying livelihoods and communities. For example, the Masai pastoralists were displaced when the Masai Mara National Reserve was established in Kenya. In Malaysia, local communities were split up and fishermen lost their livelihoods when the State Government decided to develop the island of Langkawi for tourism.

CASE STUDY: NEPAL TOURISM

Nepal is a poor, mountainous country which in 1999 had a per capita GNP of US$220. It has a population of 24.3 million which is rising by 2.5% per annum. Its biodiversity and mountain scenery provide opportunities for tourism, particularly trekking and mountaineering.

Since Nepal opened its borders to outsiders in the 1950s visitor numbers have increased. Thomas Cook offered the first organised tour to Nepal in 1955, and the first organised trek took place in 1965. Visitor numbers have risen from about 4000 to 460,000 between 1961 and 1998. There have been some fluctuations during that period and recent political instability has also threatened to undermine growth. Most tourists visit Nepal in the dry season from October to May and stay on average between 10 and 14 days. In 1996, tourism accounted for 3.8% of the GDP, generated US$117 million in foreign exchange and created an estimated 100,000 jobs. Trekking is concentrated in three locations: the Annapurna Conservation Area, which receives 60% of the visitors, around Everest in the Sagarmatha National Park, and at Langtang a remote and recently developed area (see Figure 15).

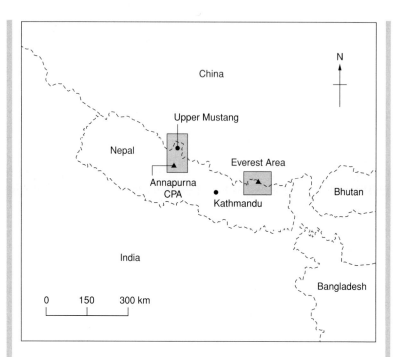

Figure 15   Map of Nepal showing trekking areas

The economic, environmental and social-cultural impacts of tourism are well illustrated in a study by Robinson of Sagarmatha (Everest) National Park. The Park was established in 1976 and became a World Heritage Site in 1980. Trekking mainly occurs in the Khumbu region, four valleys near Mount Everest. Before tourism Sherpas in the area traditionally made their living through pastoralism and trade with Tibet. Now about 85% of village households in the region earn an income from tourism. The opening up of a commercial airstrip at Lukla in the 1960s, which reduced travelling time from Kathmandu to 40 minutes, encouraged growth in tourism.

Sherpas, many of whom were subsistence farmers growing potatoes and vegetables and herding yaks, now earn a higher income by acting as porters, cooks and guides on trekking expeditions. Young men have left their villages in the Khumbu region and set up their own trekking agencies in Kathmandu. The money they earn is then sent back to the Khumbu region to help their families. In the Khumbu region itself Sherpas have converted their homes into lodges, inns and teahouses on the trekking routes. As the result of foreign loans and their own entrepreneurial approach, the Sherpas dominate lodge ownership. They use the money they earn to rebuild their own homes and to improve schools and health care. Park fees and access permits to mountain peaks also generate wealth for the area. Some income has been used to finance mini-hydro-electric schemes, which supply power to lodges thereby reducing the need to fell trees for fuel.

Positive economic benefits are offset by the fact that in villages such as Namche, food is imported into the area rather than produced locally leading to higher prices. Moreover, some villages on the trekking trails have become dependent on tourist cash rather than traditional barter. Not all Sherpas or villages in the Khumbu region have benefited from tourism, for example Thame, which is not on the trekking route, has remained poor. This has created a divide between those whose income is based on land and animal ownership and new tourist Sherpa élite whose wealth is assessed in cash. Another new trend has been the foreign ownership of accommodation, for example the Japanese Everest View Hotel near Namche now threatens Sherpa domination of the lodge industry.

In many respects tourism has not fostered a sense of cultural inferiority among Sherpas; something that often occurs in LEDC tourism. Robinson attributes this to the close-knit nature of the Sherpa community, which has helped to support host–client relations. Sherpas are respected for their resilience, rather than looked down on by trekkers. Moreover, independent 'explorer type' tourists travel in small numbers and are likely to have more empathy with their host communities than the category of visitor associated with mass tourism. Nevertheless, close contact with clients on treks, which often last for between 2 and 4 weeks, must penetrate deeply into Sherpas' personal lives. It creates the potential for change which might be greater than that associated with mass tourism.

Sherpa diet has been improved because they have been able to buy rice and eat Western food on treks. The hazardous nature of high-altitude trekking and mountaineering has, however, lead to high Sherpa fatalities on expeditions. Buddhism has not been degraded by contacts with tourists, instead the money generated has financed the rebuilding of monasteries such as that of Thyangboche. Nevertheless, the long periods spend away from villages has reduced Sherpa experience of Buddhist rituals and many have abandoned traditional robes in favour of Western dress. In the low season, Sherpas have been able to return to agriculture, but this work has fallen to women when the men are on trekking expeditions. The birth rate has also fallen because men have been away on trekking expeditions. The migration of young, able men to Kathmandu has meant that villages have lost able administrators and, as a consequence, there has been no consensus on matters such as management of common pasture and the development of new hotel facilities. Men working in Kathmandu have two homes and often two wives. Children work in the tourist industry when they should be attending school.

Changes in traditional Sherpa land-use practices had led to some deforestation in the Khumbu region, but this has considerably increased as the result of tourism. Sub-Alpine forest has been cleared for timber to build tourist lodges and upgrade Sherpa homes. Firewood has also been gathered to provide trekkers with hot food and hot showers. Clearance has led to soil erosion and landslides, and has reduced firewood supply for local people. The removal of juniper, a high-altitude and a slow-growing shrub collected for fuel, has also reduced ground cover and increased soil erosion. Park regulations allow only deadwood to be collected for fuel and tree felling requires a permit, but these regulations are difficult to enforce. Moreover this has encouraged timber and firewood collection on slopes immediately outside the park boundaries, and villages such as Monjo have become wood trade centres. Efforts to reduce the problem include: a tree nursery project at Phorte funded by the Edmund Hillary Trust, a mini-hydro-electric project at Namche; and the use of Kerosene stoves at high-altitude lodges. Afforestation is, however, a slow process and there is no motivation among locals to plant trees.

The increasing use of yaks as pack animals, rather than porters to carry tents, equipment and food, has led to increasing grazing pressure. Loss of vegetation has resulted in soil erosion. Popular trekking trails have become deeply incised and excessively widened by visitors. Local water supplies have become polluted because chemical soaps have been used for washing and toilets have been located too near streams. Beer bottles, plastic, batteries and tin cans, all of which are non-degradable, litter trails and areas around tourist lodges. These problems have increased as the Ministry of Tourism has relaxed the numbers allowed on mountaineering expeditions. The Everest base camp is said to be littered with 500 empty oxygen cylinders. Shortage of staff and funds make the task of cleaning up difficult. Concerns that degradation might have consequences for future visitor numbers have encouraged Sherpas to initiate their own clean-up campaigns supported by funds from Worldwide Fund for Nature.

There is a need for planning to control development. The area has passed through the 'exploration' and 'involvement' phases of the tourist life cycle and is now in the 'development stage'. If tourism is to progress further, management of resources will be necessary otherwise the region will lose its niche advantage. Regulation of visitor numbers entering the area is one option. Another option is to raise the relatively modest Park fees, which might discourage some visitors from coming, and use the income to fund projects that make use of alternative fuels.

The Annapurna Conservation Area Project was established in 1984 to try to ensure that trekking was managed in a more sustainable way. Solar water-heaters have been tried at some lodges to reduce pressures on firewood. Some of the income generated from tourism has been used for trail maintenance, and to set up visitor information centres to educate visitors about environmental pressures and Sherpa cultural traditions. Nevertheless, since designation, visitor numbers have increased sharply and some lodges are overcrowded.

The remote Upper Mustang region was first opened up to visitors in 1992, partly to relieve pressure on other locations. Regulations limit visitor numbers to 1000 per year; agencies must bring their own supplies of kerosene, waste must be properly disposed of and tourists must not give cash to children. Nevertheless, there have been incidences of firewood collection and tree felling, and reports of an increase in begging by children along the trails. Moreover, two video screening outlets in the village of Lo Manthang now bring Bollywood and the material world to the local Mustangis.

3 Environmental Impacts

a) Factors influencing the degree of environmental impact

The degree of environmental impact depends on a number of factors, one of which is the number of tourists involved. It is generally assumed that higher numbers cause more damage, but this is not always the case because mass tourism may bring with it greater efforts to control negative impacts. In contrast, small numbers of tourists in unregulated environments, such as hikers in the backcountry, can cause a considerable damage by felling trees for firewood, dropping litter and inappropriate waste disposal. The nature of tourist or recreational activity also influences level of impact, for example motorbike scrambling causes more damage to soils and vegetation than walking. Some locations are more vulnerable to pressure than others, for example steep mountain paths are more rapidly eroded than city pavements. Historic and prehistoric sites are also vulnerable, especially where tourists are confined in small spaces, as for example within the tombs in the Valley of the Kings in Egypt.

Without management, damage generally increases with length of exposure to visitor pressure. It can also be cumulative, for example once vegetation is removed heavy rainfall can quickly increase soil erosion on steep slopes. Tourist or recreational activities, which are seasonal in nature, allow, for example, plants on trampled footpaths to recover. On the other hand, during the high season, vegetation may be severely damaged making regeneration slow.

b) Positive environmental impacts

Emphasis on negative environmental impacts has obscured the fact that tourism can also bring benefits. Tourist revenues can be used to repair footpaths, upgrade resorts, restore historic buildings and improve run-down or polluted environments. For example, Copacabana beach, in Rio de Janeiro, in Brazil has been cleaned up for the benefit of tourism. Derelict sites in inner cities, such as the Liverpool waterfront, have been regenerated to promote urban tourism. Hotel developments can bring about improvements in water and electricity supplies for residents as well as tourists.

Natural, historical and cultural sites, which have the potential to become tourist attractions, are less likely to be lost to industrial or residential development. For example, declaring the Gower Peninsula in South Wales an Area of Outstanding National Beauty brought economic benefit to this area and protected the environment from development. Similarly, National Parks in East Africa generate valuable foreign earnings from tourism and also help to protect the wildlife from poachers. Establishing ecotourism in Kakum National Park has prevented the rainforest from timber exploitation (see case study).

CASE STUDY: KAKUM NATIONAL PARK, GHANA

Until recently this rainforest, covering 360 km^2 in central Ghana, was exploited for timber. Areas were also cleared for the cultivation of cocoa. In 1992 the Kakum Conservation Area was created to protect the park from further exploitation and to encourage sustainable tourism. The Park, which was officially opened in 1994, contains a rich variety of animals including forest elephants, rare monkeys and a large antelope called the bongo. Over 500 species of butterfly and 250 varieties of bird including hornbills, Frazer-eagle-owls and parrots also live in the forest.

A UN Development Programme donated US$3.5 million which was matched by the Ghanaian Government and NGOs to audit the parks' resources, train local forest guides, build a visitor centre and construct interpretative trails and a canopy walkway for the tourists. Park visitation has significantly increased from 2000 in 1992 to 70,000 in 1999. Visitors stay in a nearby camping area runs by locals, in hotels and lodges at Cape Coast, and further away at international hotels in Elmina. International tourists arrive at Accra airport and are then conveyed by road via Cape Coast to the park (see Figure 16). Local guides organise walks through the Park where tourists learn about the practical and medicinal uses of the forest. The canopy walkway reduces environmental pressure on the forest

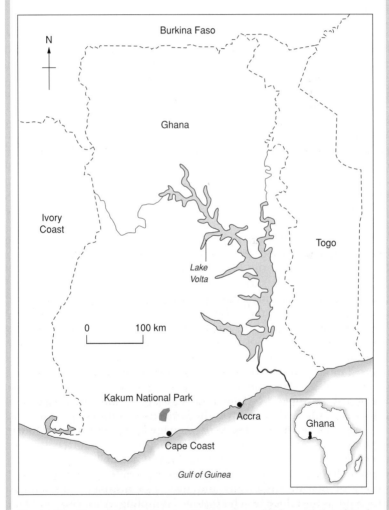

Figure 16   Map of Ghana showing the location of Kakum National Park

floor and also generates income. In the village of Mesomagor tourists visit cocoa farms, eat local Ghanaian dishes and listen to the 'Bamboo Orchestra', all of which generate employment for local people. Poaching in the park is now banned, a policy which is resented by some locals who claim they can no longer make a living from hunting or gathering foodstuffs.

c) Negative environmental impacts

Tourism can result in direct loss of agricultural and rural landscapes, damage to ecosystems, soil erosion, depletion of local resources, and noise, water, air and visual pollution. Hotel, road and airport construction destroys or fragments ecosystems to the extent that they become too small to sustain floral and faunal communities.

Deforestation, for example of slopes in the Alps to create ski runs, has caused wildlife to decline and the occurence of major snow-slides. Wetlands, for example in southern France, have been drained to create tourist developments. These habitats are important because they produce nutrients, prevent coastal erosion, filter off-land sediment and serve as important nurseries to commercial fish. On the Kenyan coast near Mombasa mangrove has been cleared for the timber trade, for fuelwood and to create aquafarms, all activities partly linked to the tourist industry.

Extraction of sand for hotel construction on Greek and Turkish beaches has disturbed the feeding and breeding patterns of the loggerhead turtle. In the Kenyan game parks sometimes as many as 30–40 minibuses may surround and disturb animals such as lions and rhinos. Guidelines concerning how near vehicles should approach animals are often ignored because drivers know they will receive extra tips from the tourists if they get in close. Animals have also been disturbed by hot air balloons, which carry tourists across the park. Elephants have died from eating zinc batteries discarded near tourist lodges. Litter attracts predator species such as foxes, which in the Cairngorms has as resulted in losses of ptarmigan and red grouse. Exotic fish have been collected for aquariums, and shells have been gathered in the Red Sea and off the Kenyan coast for the tourist trade.

Coral has been destroyed, or damaged, by a variety of tourist impacts. This organism grows in highly oxygenated, silt-free, shallow, warm seas averaging at least 25°C and is very sensitive to any environmental change in these conditions. Reefs provide habitats for a rich variety of marine organisms and they also dissipate wave energy, which helps to reduce beach erosion. Trampling by tourists, and boat anchors which drag across the surface of the coral, has resulted in damage to reefs in the Caribbean, Australia and Kenya. The practice of killing fish using dynamite has also adversely affected the coral. More indirectly, reefs have suffered because mangrove swamp has been cleared for marina and hotel construction, which has increased the amount of silt in coastal waters. Additionally, the discharge of untreated sewage into the sea has caused **eutrophication**, which in turn has led to algal growth, which has suffocated the coral. Furthermore, the Crown of Thorns Starfish, which thrives in polluted water and feeds off fish in the reef, has caused extensive damage to the Great Barrier Reef. Coral has also been collected and sold as tourist souvenirs, or made into jewellery.

Heavy trampling of footpaths in popular recreational areas has resulted in a decline in plant diversity, which in turn has adversely affected insects and insectivorous birds. Once the vegetation has been removed, rain and frost action have resulted in soil erosion and gullying on steep slopes. Paths have become deeper and wider and often multiple, as the central trench has become uncomfortable to walk in. In Kenya, minibuses carrying tourists across the game parks have created rutted roads and removed vegetation, which has led to soil erosion.

The degree of water, air and noise pollution attributed to tourist and recreational activity has often been difficult to separate out from that caused by the resident population. Some of the greatest problems have, however, occurred when mass tourism and development has proceeded at a faster rate than sewage and water supply works have been constructed (see Turkey case study). Some traditional UK seaside resorts have also encountered difficulty maintaining beach-water quality standards using sewerage systems that date from the Victorian period. Polluted sea water is aesthetically unappealing and bathers risk suffering from water-borne diseases such as gastro-enteritis and typhoid. Sea-food can also become contaminated by pollution. Around the Mediterranean it is estimated that only 30% of the sewage is treated, a factor that has contributed to the decline of some resorts. Increased agricultural activity, stimulated by additional hotel demand for local food, has led to eutrophication and algal blooms. Washed onshore during the summer months, algal blooms have been partly responsible for a downturn in the tourist economy in resorts such as Rimini on the Italian Adriatic coast. Rubbish and oil discharged from tourist boats and washed onshore by waves has contaminated beaches.

Increased air travel has raised levels of nitrous oxide, lead and hydrocarbon pollution. Pollution levels have been particularly high near airports, major road intersections and urban areas, causing respiratory problems. Alpine plants alongside the St Gotthard Pass in Switzerland, and vegetation on the floor of Yosemite National Park in California, are both reported to have been adversely affected by pollution from high vehicle flows. Tourist traffic flows have been partly responsible for sulphate weathering of the Sphinx in Cairo and the Parthenon in Athens. Noise pollution is noticeable at airports and around major roads and resorts frequented by young people, especially at night. It becomes particularly intrusive in areas where tourists seek solitude and quiet, such as the Grand Canyon, where the peace is disturbed by helicopter rides overhead. Tourist development exploits and depletes local resources, for example in Antigua, in the Caribbean, beach sand has been removed to build hotels. On the Kenyan coast, hotel demand for lobster, crab and prawns has depleted marine resources and now fish have to be imported from Tanzania. Luxury hotel complexes, especially in dry climates, make heavy demands on local water supplies. Resources are also depleted when water is used to irrigate golf courses and crops grown for tourist consumption. Local people often oppose planned tourist developments; for example Indian peasants in Tepotzlan in Mexico have protested against proposals to build golf courses, five luxury hotels and tourist villas because local water supplies will be tapped. At Taj holiday village in Goa in India the tourist demand for water has caused local water-tables to fall and wells to dry up. Locals also suffer electricity shortages because of heavy hotel needs. Hotel construction has damaged sand dunes, denied locals access to the beach and resulted in the discharge of untreated sewage into the sea. The 'Goa Foundation' has been formed to protect the interests of local people. Unfortunately, local anger has occasionally led to acts of aggression against tourists.

Tourist facilities designed or constructed in materials that do not blend in with the landscape or other buildings, create eyesores. Facilities such as ski lifts are also visually intrusive. Overcrowding at popular tourist attractions, such as the Taj Mahal in India, detract from the quality of the visitor experience.

Approaches to controlling environmental pressures include: zoning, the creation of honeypots and regulation of visitor numbers, all previously discussed in Chapter 5. Other methods include introducing planning controls, conducting environmental audits and establishing tourist codes. For example, to limit environmental pressure on the Balearic Islands the Spanish government have passed laws limiting future tourist development. The Finnish tourist board have conducted environmental audits of city hotels, farm tourism and ski resorts to encourage better methods of waste disposal and to decrease energy consumption. The German tour group Tourisk employs an environmental manager to audit all its operations and to raise hoteliers' awareness of sustainable practices. The Icelandic government has also encouraged tourist providers to conduct environmental audits (see the case study on Iceland). Other bodies have introduced voluntary codes, for example the Himalayan Tourist Code provides guidelines of expected tourist behaviour on treks. Tourism Canada issues tourist bodies with guidelines on developing sustainable practices. Dartmoor National Park has a visitor code (see Chapter 5).

CASE STUDY: TURKEY

Turkey covers an area of 780,000 km² and has 8000 km of coastline. It has a hot, dry summer climate, a rich variety of historic sites, intriguing landforms and a beautiful coast, but it was not until the 1980s that improvements in air travel made it a popular holiday destination for western Europeans. In 1963 visitor numbers were only 198,841, but by 1991 they had reached 2,398,666 and by 2001 were 11,618,969. The Turkish government was keen to encourage tourism because it would bring much-needed foreign currency earnings. It also wanted to forge closer links with Europe and to gain EU membership.

Marketed as a cheap sun–sea destination, and offered as an alternative to a beach holiday to Spain, western European visitors soon began to arrive in large numbers. Demand for accommodation outstripped supply and led to rapid, unplanned development and a number of environmental problems. Fishing villages in south-west Turkey such as Bodrum and Marmaris were transformed by haphazard hotel growth (see Figure 17). High-rise buildings grew along the coast with little regard for landforms or natural features, and some hotel sewage was discharged into the sea. Similar problems occurred at Antalya, further east. At Dalyan, extraction of beach sand for hotel construction threatened the habitat of a large population of loggerhead turtles who laid eggs on the shore. At Patava, development threatened the Mediterranean monk seal.

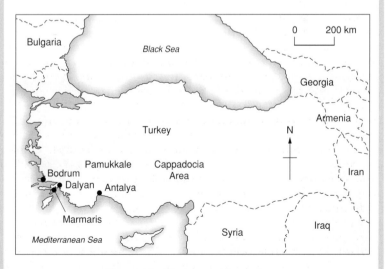

Figure 17 Turkey and its regions

Further inland, churches carved in natural rock pillars in the Cappadocia area were damaged by tourist pressure.

At Pamukkale, a dramatic calcite terrace was eroded by visitors who paddled in pools (now banned) and walked across the limestone surface (see Photo 9). Fluorine discharged from nearby swimming pools damaged the calcite and sewerage from nearby tourist developments also created problems. Hotels, by extracting thermally heated, calcite-rich water, prevented the limestone terrace from continuing to form. Moreover, in the absence of calcite-rich water, low forms of plant-life had begun to establish themselves. The development of stalls selling cheap souvenirs along the rim of the terrace also created a visual eyesore.

A number of steps have been taken to reduce environmental impacts. At Pammukkale, UNESCO offered money to remove the cheaper hotels and stalls and to restore the Royal Pool so that fluorine was no longer discharged. The turtles on Dalyan beach are now protected and have become a tourist attraction. The government, aware of the need to safeguard the tourist industry, set up the South West Environmental Project to reduce inappropriate wastewater disposal and prevent future pollution along a 2000 km stretch of the Aegean and Mediterranean coastline. The government is keen to diversify the tourist industry, and spread the benefits of tourism across a greater proportion of the country. One plan is to expand winter sport facilities, but also at the same time ensure that developments are carefully planned. Another scheme involves reviving the Silk Road, an ancient trading route, by restoring caravanserais (Eastern inns with an inner court) along its path. Attempts will also be made to improve tourism in the Black Sea region.

Photo 9   Tourists on the calcite terrace at Pamukkale

CASE STUDY: ICELAND

Iceland covers an area of 103,000 km². Only the southern coastal fringes are suitable for agriculture. Most of the island is composed of bare rock and gravel surfaces overlain by glaciers. The population in 2002 was 288,200, two-thirds of whom lived in Reykjavík.

Iceland's spectacular waterfalls, glaciers, geysers, coastal scenery and wildlife have given the country a rich natural resource base for the development of tourism. It also has a strong cultural heritage associated with the Icelandic sagas and the Vikings. Improvements in air travel and active promotion by the government and travel companies have turned Iceland's resources into tourist attractions. A curiosity to see and learn about natural environments, together with a desire to engage in active forms of recreation such as hiking, horse-riding and snowmobiling, also explains why Iceland has grown as a tourist destination.

As a consequence of the aforementioned factors, visitor numbers doubled in the 1990s, reaching just over 300,000 by 2000. Growth is reflected in an increase in hotel accommodation, improvements to the international air terminal at Keflivík, the siting of interpretation boards at points of interest around the island, and new visitor centres in Reykjavík and Thingvellir National Park.

It is estimated that tourism has created 5400 jobs and now contributes 4.5% of the GNP and generates 13% of all Iceland's foreign earnings. Almost all travellers arrive by air and stay for about 10 days in summer and 5 in the winter. About 80% of visitors come for a holiday, while a further 11% travel for business reasons. Iceland is particularly popular with visitors from Norway, Sweden, Denmark, the USA, Canada and the UK.

Typically short-stay visitors make Reyjavík their base from where they travel to see Thingvellir National Park in the rift valley; the hot springs at Geysir and Gullfoss waterfall (see Figure 18). They may also journey along the south coast to Vik. Visitors staying for a longer period often make a circular tour of the island to see Vatnajökull glacier, Dettifoss waterfall and Húsavík on the north coast from where they can go whale watching. Most tourists swim in some of the numerous geothermal pools during their stay.

Rising visitor numbers has invariably raised concerns about environmental pressure on sensitive sites, particularly because cool temperatures limit plant growth and rain causes erosion on bare slopes. Paths near popular waterfalls such as Dettifoss have been eroded. Some paths, for example at Gullfoss, have been repaired using materials and designs that blend in with the landscape. Driving on unmarked roads is prohibited to limit damage to vegetation. Several Icelandic travel companies, aware of the need to develop sustainable tourism, have had their businesses officially certified as ecofriendly. For example, 'Icelandic Farm Holidays' has become a member of 'Green Globe 21' and the Blue Lagoon geothermal pool has been awarded a Blue Flag certificate in recognition of the work on environmental issues. Thingvellir, the first national park to be established in 1928, has been nominated as a UNESCO site. The most recently designated national park is Snæfellsjökull created in 2001.

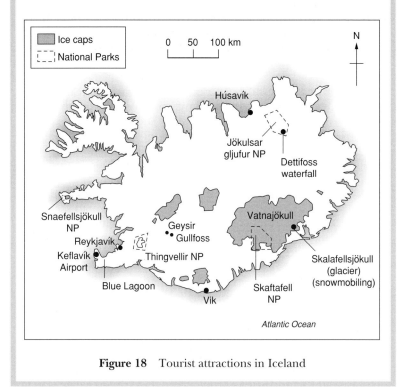

Figure 18 Tourist attractions in Iceland

Summary

1. Tourism generates positive and negative economic, socio-cultural and environmental impacts.

2. Level of socio-cultural and environmental tourist impact is dependent on the number of visitors involved, the nature of their activities and length of exposure to outside influences or pressures.
3. Zoning, the creation of honeypots, regulation of visitor numbers, planning controls, tourist codes, environmental audits and dispersal of visitors across the attraction can alleviate environmental pressures.

Questions

1. What do you understand by the terms: (a) travel account, (b) leakage, (c) commodification?
2. Produce a concept map showing the negative and positive environmental impacts of tourism.
3. Tourism in any location is likely to generate as many costs as benefits. Discuss this view with reference to case studies.
4. Suggest strategies for reducing environmental pressure on popular tourist locations.
5. Even if LEDCs have a perfect climate and beautiful beaches they should not rely solely on tourism for development. Discuss.

7 Sustainable Tourism and Recreation

1 Characteristics of Sustainable Tourism

Sustainability was defined by the World Commission on Environment and Development in 1987 as 'development, which meets the needs of the present without compromising the ability of future generations to meet their own needs'. It therefore follows that sustainable tourism should not consume renewable resources faster than rates of natural replacement. **Sustainable tourism** is not a preservation of the physical environment, but development based on economic, socio-cultural and environmentally sustainable principles. Economic sustainability involves encouraging continued investment in tourism and maintenance of local jobs over time. Projects involve local people, and the economic costs and benefits are distributed between tourist developers and hosts. Socio-cultural sustainability protects and enhances local customs, livelihoods and cultures. Environmental sustainability ensures that natural landscapes and ecosystems are conserved or enhanced by tourist developments. Rising visitor numbers should not detract from the quality of the tourist experience and environmental carrying capacities must not be exceeded. Planning controls should be in place, before building occurs, to ensure that the scale, nature and character of the tourist facility are suited to, and blend in with, the environment. Changes to the infrastructure should integrate with regional policy so that other sectors of the economy can benefit from improved linkages while also ensuring that natural environments are protected.

In reality, sustainable tourism can be difficult to achieve. Local communities may instead favour other economic options. For example, when the Rio Tinto Zinc Company (RTZ) suggested mining for titanium dioxide in the Natal province in South Africa, an area of coral reefs, grassland swamps, afforested dunes and deserted beaches, the government opposed the plan on environmental grounds and instead proposed developing the area for ecotourism. The local Zulus, however, favoured the mining option because RTZ had a good track record of paying high wages and in investing in local schools, clinics and hospitals. In contrast, the Natal Park Board, which managed surrounding parks, were perceived by the local people as a body who paid low wages and displaced communities from land in order to establish game reserves.

Plans to build a funicular railway in the Cairngorm ski area in Scotland in the 1990s were opposed by the Worldwide Fund for Nature and the Royal Society for the Protection of Birds on the grounds that it would damage fragile alpine plant communities. Local people, however, welcomed the proposal because it would bring economic benefits to an area where there were few other types of employment available.

From these principles it therefore follows that sustainable tourist developments generally involve small numbers of tourists, are small scale and dispersed in nature. Sympathetic design and materials are used in the construction of tourist facilities. Host communities are involved in, or take control of, developments leading to improvements in their local economy. Capital and running costs are relatively low. Developments protect and/or enhance existing natural environments, cultures and traditions.

Sustainable tourism is often linked to **alternative tourism**. This type of tourism generally involves small numbers of tourists and is small scale and dispersed in nature. Sympathetic designs and materials are used in the construction of tourist facilities and host communities are involved in, or take control of, developments, which lead to improvements in their economy. Capital and running costs are relatively low. Developments protect and/or enhance existing natural environments, cultures and traditions, for example the Kapawi Eco Lodge in Ecuador.

A common criticism of alternative tourist projects is that they generate insufficient jobs to benefit the national economy. For this reason, a combination of sustainable tourism and controlled mass tourism is seen by some as the best way forward. Examples of this approach can be found in Sri Lanka where tourist enclaves occur around the coast, while smaller-scale, alternative sustainable projects are found inland.

Although sustainability is often equated with alternative tourism, efforts are being undertaken to make mass tourism more 'green' through, for example, more efficient use of water and energy and the recycling of waste. Management of wilderness areas in many MEDCs also embraces sustainable practices (see Kakadu National Park case study here and also other case studies in Chapter 5). The term '**green tourism**' refers to all forms of tourism that attempt to reduce environmental and ecological impacts.

2 Measuring Sustainability

There have been a number of attempts to evaluate whether tourism is sustainable, one of which is to assess carrying capacity (see Chapter 4). Such a method does, however, have limitations, for example environmental carrying capacity is difficult to anticipate and tends to focus on damage to soils and vegetation rather than other impacts such as noise pollution. An increase in car park size can modify physical carrying capacity. Assessments of perceptual carrying capacity are very subjective because people vary in the level of overcrowding they can tolerate. Attitudes to overcrowding also depend on the nature of the recreational activity in which people participate.

An alternative method of evaluation, used in USA National Parks, is Limits of Acceptable Change (LAC). This technique recognises that change is inevitable as tourism develops but tries to ensure that planning controls are in place to manage growth. Specifically, it ensures that there is an agreed criterion for development, all interested parties are represented in the decision-making, and that proposed changes after development are carefully monitored and adhere to the conditions laid down at the planning stage. One problem with LAC is that of agreeing and assessing qualitative aspects of tourist development. Another difficulty is that the planning process requires resources, capital and expertise, which are expensive. For is reason LAC is perceived as impractical for many LEDCs, for example The Gambia.

A further approach is to undertake an Environmental Impact Assessment (EIA). Using a variety of data sources including maps, EIAs identify the activities and induced activities that planned tourist developments such as marinas produce. They highlight the direct and indirect environment impacts of any proposed development and suggest strategies to minimise these while also maximising the benefits of the project. Criticisms of EIA are that too much emphasis is placed on environmental rather than wider impacts, it is costly to carry out and tends to be used only on large-scale projects.

3 Ecotourism

Ecotourism has developed in the last 20 years, having risen sharply from a very small base, and now accounts for 5% of the tourist market. Often perceived as the opposite of mass tourism, ecotourism tends to occur in remote, exotic locations such as the Galapagos Islands and

Antarctica. Holidays tend to be expensive and run by independent specialist operators who cater for relatively small numbers. Visitors undertake travel to ecotourist destinations to learn and understand about landscapes and ecosystems that differ markedly from their own. Ecotourism embodies sustainable practices and involves local people who gain financially and use the revenues to conserve their natural resources. Contacts between visitors and hosts aim to encourage both groups to appreciate the need to conserve and manage tourism in a sustainable way.

Ecotourism grew as a reaction to some of the criticisms of mass tourism, namely the cultural debasing of host communities and the transfer of tourist revenues from poor to rich countries. Other incentives for growth have been that beach holidays have become less fashionable and that improvements in education have made people more curious to visit more remote locations with a rich cultural, historic or biological value. Furthermore, improvements in education, and gap-year or post-university experiences, have given travellers more confidence in interacting with multi-cultural groups. Travel companies have been quick to respond to changes in public taste and have opened up new locations. Television, advertising and Internet marketing have helped to promote packages that are increasingly tailored to individual needs, for example the Galapagos Islands.

One difficulty with ecotourism is that small visitor numbers can restrict opportunities to benefit from economies of scale. Few local jobs are also created which in turn reduces any multiplier effects. Ecotourism also tends to occur in fragile, sensitive environments such as coral reefs, which can support only low capacities before damage occurs. LEDCs may also lack the expertise to cater for tourists, which means they use inappropriate practices and local resources are exploited in the short term. The high cost of ecotourism holidays has led to claims of élitism and the suggestion that the 'eco-label' is used by some travel firms as a ploy rather than genuine policy of responsible tourism (see Belize case study). There are also concerns that the greater contacts developed between tourists and locals may mean that Western culture penetrates deeper into their personal lives than some forms of mass tourism.

CASE STUDY: KAKADU NATIONAL PARK, AUSTRALIA

Kakadu National Park is located about 120 km east of Darwin in Australia's Northern Territory in the Alligator River region (see Figure 19). Covering an area of 19,804 km² the area is home to 75 species of reptiles including frill-necked lizards and salt-water crocodiles. Scenery is diverse and includes sandstone plateaus and escarpments, savanna woodlands, floodplains, mangroves

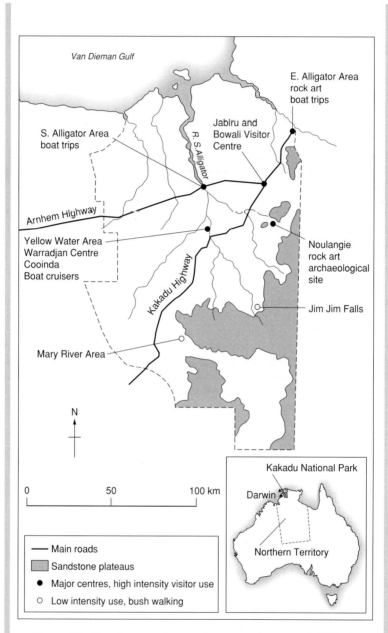

Figure 19 Tourist attractions and conservation areas in Kakadu National Park, Australia

and mudflats. The area is also rich in Aboriginal art and archae-ological sites. The whole park was designated a World Heritage Site in 1992.

Land in the park is owned by Aborigines who lease it to Parks Australia. The Park is jointly managed by the Aboriginal owners and Park Directorate. The objectives of tourist management are to provide visitors with enriching experiences, to protect the natural and cultural heritage and to ensure that the Bininj/Mungguy Aborigines benefit from tourism.

Kakadu has grown in popularity as a tourist destination and attracts on average 300,000 visitors per year. Most tourists come during the dry season between May and October, about one-third are on organised tours and their average length of stay is between 2 and 6 days. Visitor facilities within the Park include: the Bowali Visitor Centre, the Warradjan Aboriginal Cultural Centre, walk-ways, viewing platforms and interpretation panels at the rock art sites, camping grounds, bush camping sites, picnic areas, boat ramps on the lake shore and bird hides. The main settlements within the Park are Jabiru and Cooinda, which have hotels and hos-tels and an airstrip from where visitors can depart on sightseeing tours over the Park. Applications for a jet standard runway have so far been rejected by the Park authority. Kakadu Park is promoted by the Northern Territory Tourist Commission and Parks Australia.

About 300 Aborigines live in the park and tourism has benefited them in a number of ways. Park status guarantees protection of their sacred sites and their Aboriginal rights, and they receive money from the lease. Some Aborigines own hotels and hostels while others are employed as rangers.

The Park has been zoned into a number of areas, which include localities where tourists have access, zones set aside for scientific research and conservation, and areas for Aboriginal use only. About one-third of the park is wilderness and here driving is inappropriate. Commercial vehicles using the Park must have a permit and a visitor code reminds tourists not to feed animals or drop litter.

The tourist zones themselves are further subdivided into areas of high- and low-intensity use. High-intensity areas coincide with scenic drives, designated campsites and visitor centres. At these locations surfaces are hardened to withstand high visitor press-ure, campsites have hot and cold water and flushing toilets, and picnic sites are close to vehicle access. In low-intensity areas vehi-cle access is limited and bush camping requires permits. Further restrictions at low-intensity areas may be introduced in the future if visitor pressure increases. Options include imposing weight restrictions on vehicles, limiting visitor numbers or closing areas to commercial vehicles.

CASE STUDY: BELIZE

Belize in Central America, formally known as British Honduras, has witnessed a large growth in tourism in recent years. Numbers of international arrivals rose from 88,000 to 129,000 between 1984 and 1994. It has become a popular holiday destination, especially for 'ecotourists' because it is politically stable, English speaking, is just 2 hours' flying time from Miami, possesses many Mayan archaeological sites and has a rich biodiversity. Its ecosystems include mangrove swamp, savannah, pine forest and a coral reef barrier said to be the second longest in the world. It has a number of reserves including the Cockscomb Basin and Cheetah Sanctuary where forest once exploited for cedar and mahogany is now preserved. Visitors to Cockscomb Basin can see a variety of animals and birds including black howler monkeys, tapir, keel-billed toucans and scarlet macaws (see Figure 20).

One example of a successful community ecotourism project is the Baboon Sanctuary, a reserve set up by local villagers and the WWF to protect black howler monkeys. These animals have become a tourist attraction and their numbers are now increasing. Local people derive income from taking tourists on guided walks through the reserve. In another development, the Toledo Ecotourist Association (TEA), supported by the Belize government, helped to construct guesthouses in six villages in the district of Toledo in southern Belize. The accommodation, which is managed by local people, is built in traditional styles and uses local materials. Each guesthouse sleeps eight people and the water supply is provided from simple tanks.

Such schemes have created local employment but there have also been difficulties. For example, hotel and lodging house proprietors in district capital Punta Gorda were opposed to the TEA project because they feared it would threaten their businesses. Another problem has been that although TEA recognised that tourists would need to be rotated around the villages to ensure a fair distribution of benefits, in reality villages have competed for tourists. The problem has been made worse because a charity called USAID, acting through an agency called BEST (Belize Enterprises for Sustainable Technology), also operates community-based ecotourism in the same area but encourages competitiveness. One consequence of this is that a new guesthouse in Laguna built by BEST now competes with an established TEA guesthouse also in the village.

The most popular holiday destination in Belize is Ambergris Caye, an island inside the Barrier Reef. The only settlement on the island is San Pedro where the first hotel appeared in 1965. Hotels now exceed 30 and visitor numbers are expected to rise

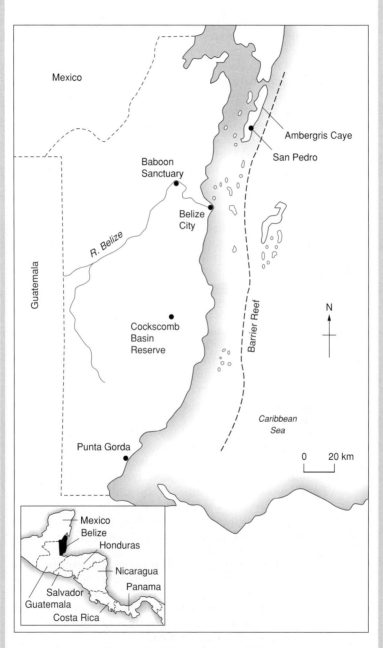

Figure 20   Tourist areas in Belize

to between 70,000 and 90,000 by 2005. In terms of the 'Butler resort model' San Pedro has moved from a situation where a few 'explorer' type visitors stayed in a smaller number of locally owned hotels to a stage where large numbers of tourists are accommodated in big, foreign-owned hotels and condominiums. Hotel and lodging developments, golf courses and residential homes now occupy two-thirds of Ambergris Caye. Adjacent to the island is Hol Chan Marine Reserve, which occupies 1 km of the Barrier Reef. The reserve was created in response to concerns about tourist pressure.

Tourism has created a lot of local employment, for example in hotels, although foreign workers hold more responsible positions. Other local people work as taxi drivers, guides or in shops and restaurants. Better pay and conditions has tempted fishermen to work as tour or dive guides. Others continue to fish when stocks are good and sell their catches directly to hotels rather than cooperatives, which give them more profit. Tourism has also created employment in the construction industry. All these changes have meant that many local people now have a higher standard of living than they did before tourism arrived. They also have access to better transport and a greater range of goods in the shops.

These positive impacts have been offset by the fact that the price of land, especially on the beach front, has increased which has meant that locals can no longer afford to buy here, which has created friction. Hotel complexes tend to overshadow other dwellings and local people feel they have had little say in the planning process particularly because the US-aided Belize Tourist Board is well represented by expatriates. The construction of hotel perimeter walls and sea-walls has increased beach erosion. This process has also been encouraged by the practice of removing grass from the beach to make it more attractive for tourists. Rising visitor numbers created more rubbish to be disposed of and has also led to more air, soil and groundwater pollution. Not all sewerage has been adequately treated before being discharged into the ocean. High demand for water from tourist hotels has caused groundwater supplies to fall.

Some economic leakage has occurred because foreign-owned hotels have purchased imported food supplies. The construction industry has also imported materials, especially complex equipment such as air-conditioning. Young local people have been particularly susceptible to negative 'demonstration effects' and some dive guides now spend their free time in bars after work rather than at home with their families.

The boom in hotel and condominium construction encouraged low-paid migrant workers from the mainland to move into

San Pedro in search of work. However, little affordable accommodation was available in the town, which has led to the construction of unplanned, shanty developments on infilled mangrove swamps.

Visitors have damaged the Barrier Reef and black coral has also been collected and sold for jewellery in San Pedro. Lobster and whelk catches have declined as local fishermen have sought to supply tourist hotels. However, a proposed tourism complex to be built on mangrove swamp near Hol Chan Marine Reserve was prevented as the result of opposition by environmental groups.

In conclusion, Belize has used the 'ecolabel' to promote tourism, but many of the developments have fallen short of good practice. The problems are, however, complex because small-scale sustainable projects cannot provide the economic returns to improve the infrastructure and services, which Belize desires. Successful tourist developments also require a greater understanding of how the needs of indigenous people, multi-nationals, aid agencies and government are interrelated; needs should not be considered in isolation.

4 Mass and Alternative Tourism – the Future

The future of mass tourism is controversial. Some researchers believe that new technologies, which have created artificial ski slopes and indoor leisure centres, some with pools that can generate wave effects, will help to undermine mass tourism which by its nature is a highly seasonal industry. Moreover negative environmental impacts of mass tourism such as overcrowding, litter, and air, water and noise pollution will hasten its downfall. Others argue that mass tourism will continue to grow because there are potential customers who have yet to enter the market, or who are about to, for example from Central and Eastern Europe. Additionally, the idea of taking a second or a third holiday is spreading to the middle and lower classes. Furthermore, there are still many other mass tourist destinations to develop as more established resorts start to decline.

The future of alternative tourism is also problematic. The high cost of alternative tourism will continue to preclude the majority of tourists from engaging in this form of activity. Others may have no desire to visit remote locations and sample the kind of experiences on offer, in which case more effort will be needed to make mass tourism greener. Conversely, 'alternative tourism' may grow in popularity, which raises concerns that in time this may lead to a new form of mass tourism, which will pass through a boom and bust life cycle.

Summary

1. Both alternative and mass tourism can embrace sustainable practices.
2. Measures to assess sustainability include carrying capacities, limits of acceptable change and an environmental impact assessment.
3. Ecotourism has grown significantly in the last 20 years. Critics of ecotourism argue that small visitor numbers produce few local jobs and little opportunity to benefit from economies of scale. Ecotourism is also perceived as élitist and travel companies use the label as a ploy.

Questions

1. Draw a diagram to summarise the main differences between mass and alternative tourism.
2. Suggest methods of assessing sustainability.
3. What is sustainable tourism? With reference to case studies examine the attempts by governments and other bodies to achieve greater sustainability in tourism and recreation.
4. Using examples, explain how and why governments and other bodies have promoted ecotourism.
5. Research a sustainable, or ecotourism, project using web searches and other resources. Good examples can be found in Ecuador and Costa Rica. Draw a map showing the location of the project. Find out about visitor numbers, attractions and evaluate sustainable practices. Assess the impact of the project on the local economy, culture and environment.

Bibliography

Badger, A. *et al.*, 1996 *Trading Places. Tourism as Trade* (Wimbledon, London: Tourism Concern).

Boniface, B.G. and Cooper, C.P., 2001 *The Geography of Travel and Tourism* (London: Heinemann Educational).

Briguglio, L. *et al.* (eds), 1996 *Sustainable Tourism in Islands and Small States: Case Studies* (London: Pinter).

Butler, R. and Hinch. T., 1996 *Tourism and Indigenous Peoples* (London: International Thomson Business Press).

Franklin, A., 2003 *Tourism* (London: Sage).

France, L., 1997 *Sustainable Tourism* (London: Earthscan).

Law, C.M. (ed.), 1996 *Tourism in Major Cities* (London: International Thomson Business Press).

Hall, C.M. and Law, A.A., 1998 *Sustainable Tourism* (Harlow: Longman).

Hall, C.M. and Johnson, M.E., 1995 *Polar Tourism* (Chichester: Wiley).

Harrison, L.C. and Husbands, W. (eds), 1996 *Practicing Responsible Tourism* (Chichester: Wiley).

Holden, A., 2000 *Environment and Tourism* (London: Routledge).

Montanari, A. and Williams, A.M., 1995 *European Tourism* (Chichester: Wiley).

Prosser, R., 2000 *Leisure, Recreation and Tourism* (London: Collins Educational).

Ringer, G. (ed.), 1998 *Destinations: Cultural Landscapes of Tourism* (London: Routledge).

Seaton, A.V. (ed.), 1994 *Tourism: The State of the Art* (Chichester: Wiley).

Shaw, G. and Williams, A.M. (eds), 1997 *The Rise and Fall of the Seaside Resort* (London: Mansell).

Timothy, D.J. and Boyd, S.W., 2003 *Heritage Tourism* (Harlow: Pearson Education).

Towner, J., 1996 *A Historical Geography of Recreation and Tourism in the Western World 1540–1940* (Chichester: Wiley).

Shaw, G. and Williams, A., 2002 *Critical Issues in Tourism* (Oxford: Blackwell).

Warn, S., 1999 *Recreation and Tourism* (London: Nelson).

Williams, S., 1998 *Tourism Geography* (London: Routledge).

Williams, S., 2003 *Tourism and Recreation* (Harlow: Pearson Education).

Van den Berg, L. *et al.*, 1995 *Urban Tourism* (Aldershot: Avebury).

Vellas, F. and Bécherel, L., 1995 *International Tourism* (Basingstoke: Macmillian Press).

References to models mentioned in text

Barrett, J.A., 1958 *The seaside resort towns of England and Wales*, unpubl. PhD thesis (London: University of London).

Butler, R.W., 1980 'The concept of a tourist area cycle of evolution, implication for management of resources'. *Canadian Geographer*, Vol. 24, No. 1, pp. 5–12.

Smith, R.A., 1991 'Beach resorts; a model of development evolution', *Landscape and Urban Planning*, Vol. 21, No. 3, pp. 189–210.

Periodicals

Geography Review, The Geographical Magazine, Geography, Transactions of Institute of British Geographers. All occasionally carry articles on tourism.
Journals devoted to exclusively to research on tourism and recreation include: *Annuals of Tourism Research* (Elsevier), *International Journal of Tourism Research* (Wiley), *Tourism Management* (Butterworth-Heineman), *Tourist Studies* (Sage).

Websites

There is a vast amount of information on the Internet on tourism and recreation. Many sites are commercial dedicated to selling holiday packages but they also contain information on attractions, facilities and maps.
Particularly useful sites are:

www.visitbritain.com – this organisation was established in 2003 when the English Tourism Council and British Tourist Authority merged, it contains useful statistics.

www.world-tourism.org – the world tourism authority website provides data on information on the latest global trends and impacts of events on tourism.

www.english-heritage.org.uk

www.nationaltrust.org.uk

www.tourismconcern.org.uk – forum for concern on tourism impacts.

Index